Electronic Gadgets for the Evil Genius™

Evil Genius™ Series

Arduino + Android Projects for the Evil Genius

Bike, Scooter, and Chopper Projects for the Evil Genius

Bionics for the Evil Genius: 25 Build-It-Yourself Projects

Electronic Circuits for the Evil Genius, Second Edition: 64 Lessons with Projects

Electronic Gadgets for the Evil Genius, Second Edition

Electronic Gadgets for the Evil Genius: 28 Build-It-Yourself Projects

Electronic Sensors for the Evil Genius: 54 Electrifying Projects

15 Dangerously Mad Projects for the Evil Genius

50 Awesome Auto Projects for the Evil Genius

50 Green Projects for the Evil Genius

50 Model Rocket Projects for the Evil Genius

51 High-Tech Practical Jokes for the Evil Genius

46 Science Fair Projects for the Evil Genius

Fuel Cell Projects for the Evil Genius

Holography Projects for the Evil Genius

Mechatronics for the Evil Genius: 25 Build-It-Yourself Projects

Mind Performance Projects for the Evil Genius: 19 Brain-Bending Bio Hacks

MORE Electronic Gadgets for the Evil Genius: 40 NEW Build-It-Yourself Projects

101 Outer Space Projects for the Evil Genius

101 Spy Gadgets for the Evil Genius, Second Edition

123 PIC® Microcontroller Experiments for the Evil Genius

123 Robotics Experiments for the Evil Genius

125 Physics Projects for the Evil Genius

PC Mods for the Evil Genius: 25 Custom Builds to Turbocharge Your Computer

PICAXE Microcontroller Projects for the Evil Genius

Programming Video Games for the Evil Genius

Raspberry Pi Projects for the Evil Genius

Recycling Projects for the Evil Genius

Solar Energy Projects for the Evil Genius

Telephone Projects for the Evil Genius

30 Arduino Projects for the Evil Genius, Second Edition

tinyAVR Microcontroller Projects for the Evil Genius

22 Radio and Receiver Projects for the Evil Genius

25 Home Automation Projects for the Evil Genius

Electronic Gadgets for the Evil Genius™

Bob Iannini

Second Edition

McGraw Hill Education

New York   Chicago   San Francisco   Athens   London
Madrid   Mexico City   Milan   New Delhi
Singapore   Sydney   Toronto

Sponsoring Editor
Judy Bass

Editing Supervisor
Stephen M. Smith

Production Supervisor
Pamela A. Pelton

Acquisitions Coordinator
Amy Stonebraker

Project Manager
Yashmita Hota,
Cenveo® Publisher Services

Copy Editor
James K. Madru

Proofreader
Avinash Kaur and Hardik Popli,
Cenveo Publisher Services

Indexer
Arc Films, Inc.

Art Director, Cover
Jeff Weeks

Composition
Cenveo Publisher Services

About the Author

Bob Iannini runs Information Unlimited, a firm dedicated to the experimenter and technology enthusiast. Founded in 1974, the company holds many patents, ranging from weapons advances to children's toys. Mr. Iannini is the author of *Electronic Gadgets for the Evil Genius: 28 Build-It-Yourself Projects* and *MORE Electronic Gadgets for the Evil Genius.*

Contents

Contents

Introduction

Full-color, full-size, high-resolution versions of all figures are available for free download at www.amazing1.com/eg3.

If any clarification or greater detail is needed for any of the images in this book, these online pictures should provide it for you.

Corrections will also be posted on this website for any chapters requiring clarification, so be sure to visit it before beginning fabrication.

Parts and Components

Most parts can be obtained at well-equipped local electronics hobbyist stores, or online at websites like Mouser or DigiKey. Some of these projects may have a few expensive components (unfortunately this is unavoidable, as these high-performance, high-power components have substantial material requirements) that can sometimes be found at reduced prices in the used or secondary markets. A few components are the proprietary design of Information Unlimited, as either these parts did not exist at all, or finding a reliable source proved exceedingly difficult. So, they are manufactured directly and available at www.amazing1.com for a nominal cost. You may still be able to find them from other sources through diligent search, or the experienced fabricator may be able to build them in-house. If any components prove difficult to find, all parts are available at this website individually, or as complete kits to be assembled (a full build of the device is still required; see each chapter's "Difficulty" and "Tools" sections for assembly overview).

CHAPTER 1

Automatic Programmable Charger

Overview

This device charges high-energy banks of electrolytic, photoflash, and other storage capacitors from 500 to 10,000 V (note that higher voltages up to 25 kV are possible—go to www.amazing1.com for additional data). It is an adjustable voltage source with short-circuit protection. Recommended capacities are between 100 and 10,000 μF. This equates to many thousands of joules! (For comparison, the kinetic energy of a .30-06 rifle is 750 J.)

Hazards

Charges capacitors to the dangerous high voltages are necessary for powering effective kinetic devices. High-voltage handling skills are required for charging over 50 J of energy. Eye protection should be worn when making, testing, and operating this device.

Difficulty

This project requires intermediate skills in wiring and soldering, along with certain basic sheet-metal work to form and fabricate sheet-metal chassis, covers, and so on. Or you may be able to purchase a chassis/electrical box with close enough dimensions from electronics supply houses as Mouser or Digi-Key, at which point all you need to do is drill the holes.

Tools

Basic wiring, soldering, hand tools, bending apparatus, voltmeter capable of measuring to 10 kV, and a low-cost oscilloscope. Electronic ability to use test equipment such as meters and a scope is necessary in completing this project.

You should familiarize yourself with Chap. 21 before building this circuit or any similar circuit without an isolation transformer. It is advised that you use the testing circuit jig, as shown in Chap. 21, for this project because it verifies dangerous alternating-current (ac) grounds. Other amenities are powering a virgin circuit starting with low input voltage and slowly increasing, noting that any excessive current could be dangerous and totally wipe out your hard work!

Use

- The controlled charging of energy-storage capacitors, such as those used for other projects in this book

- Automatic programmable high-power charger from 500 to 10,000 V for electrolytic, photo flash, high-energy storage, and other equivalent capacitors and combinations

- Optional designs available at www.amazing1 .com for greater or more specific voltage and current values

- Excellent for capacitor integrity testing, charging medium-sized energy-storage banks

1

for rail, coil, electromagnets, plasma guns, wire exploding, magnetic shaping, and so on or just as a charger for laboratory functions and experiments

■ May also be used as a current source power supply.

This unit is manually voltage-controlled by a front-panel potentiometer (voltage-level dial S2; see Fig. 1-29). The front-panel meter displays the charging voltage, helping to prevent overcharging and potentially dangerous explosions. Charging is current-controlled by the unique circuitry and does not require a power-robbing resistor. The unit operates from direct 115-Vac power. The charging rate is over 200 J/s but will depend on load impedance. Assembled size is $10 \times 7 \times 3\frac{3}{4}$ inches.

The advantages of the current source are realized in the charging of capacitors used for high-energy storage, single-event discharges intended for rail guns, and other electrokinetic and electromagnetic pulse (EMP) devices. Even at the time the charging cycle starts, the current is controlled and remains relatively constant throughout the charging cycle.

The source also can be short-circuited, and the current is still in a controlled mode. The charge voltage can be determined by a simple formula:

$$v(t) = \frac{f(t) * I}{C}$$

This is a linear relationship and does not require the power-wasting series resistance used for voltage sources. The disadvantage is that there is a tendency for the voltage to soar above the design ratings of the capacitor or part used. Operation requires preventive measures such as monitoring the charge voltage or using an electronic trigger that shuts the source off when the desired charge voltage is reached.

The advantage of the voltage-source approach is the property of maintaining a constant voltage value within the design limits required as current (load) is being drawn. This is an advantage for electrical-field charging when used for accelerators, ion beams, isotope enrichment, particle beams, potential generators, and so on. The disadvantage is that current is not controlled, and overloads can damage the circuitry. Charging capacitors now

Figure 1-1   Assembled charger.

require a power-robbing series resistance to control the current. Charging is exponential.

The HVOLTV high-power series uses a bit of both worlds, being voltage and current sources by user selection.

Special Note on Charging Capacitors

DANGER Charged capacitors can kill by electrocution and cause very serious injury from burns and explosions. Do not attempt to use this equipment unless completely educated in high-voltage circuitry. These products are intended for training and for experienced laboratory personnel.

Programmable capacitor chargers allow you to dial in the charging voltage and view the event via a front-panel analog meter. Capacitors charge from a current source and waste little energy, such as that when using a ballast resistor. Ballasted energy is all reactive and is never consumed as real power.

A selector switch is included allowing the charge cycle to cease once the preprogrammed value is reached. It also can be selected to automatically recycle the charging function, keeping the capacitor up to voltage until the charger is turned off or the event occurs.

The charging time can be estimated by calculating the energy in joules of the capacitor being charged and dividing by joules (watt-seconds) of the charger. This is only approximate because it assumes linear power being applied throughout the charge cycle.

A charge current control allows adjustment of charging time and is very useful when using smaller capacitance values.

The charging time of the capacitors can be approximated more accurately by using the simple formula

$$t = \frac{cv}{i}$$

where t = seconds, c = capacity in farads, v = voltage, and i = charging current in amps.

Brief Project Description of Functions

A high-power electronic circuit charges capacitors from 500 to 10,000 V at recommended capacities from 0.01 μF and upward. *Example:* Charging a 1000-μF capacitor to 5000 V equates to over 12,500 J.

Units are voltage-programmable by the setting of the charge-voltage control on the front panel. A meter monitors the charging voltage as it builds up to the preset value and then stops. A push button starts the charging event. Another push button stops the charging function at any time.

A charge-rate control allows adjustment of charging current/charging time and is especially useful when charging smaller capacitance values.

A selector switch allows manual or automatic charging. Manual charging ceases once the preset voltage is reached. Automatic charging keeps the capacitor indefinitely at the preset voltage; capacitors automatically charge to and maintain this voltage until discharged. A safety lamp displays any charge over 70 to 90 V (the exact amount varies with temperature) that remains on the capacitor.

Charging is lossless and does not use a resistor for current control. It is controlled by the unique circuitry, with voltage indicated by a large analog panel meter.

Connections are made via banana jacks and cables as shown.

DANGER A serious deadly shock hazard exists when using with high-energy capacitors above 50 J.

Circuit Description

The CHARGESERIES converts 120-Vac line voltage into a variable direct-current (dc) voltage to charge a target capacitor. This is accomplished

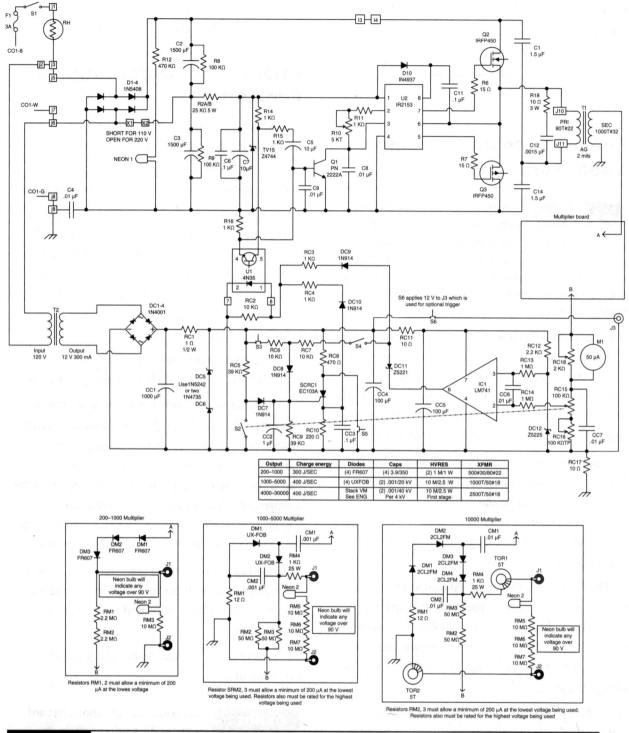

Figure 1-2   Charger schematic.

by a half-bridge inverter, in which the rectified line voltage is converted into a high-frequency wave.

The direct ac line voltage is rectified and voltage-doubled by diodes (D1,3) and accumulated by filter capacitors (C2,3). The 120 Vac is now converted to $120 \times 2 \times 1.4$ = approximately 340 Vdc. The 1.4 is the peak value because the input is capacitive to the filter/doubler network. While this system lacks regulation, it does allow more power to be output. In-rush current is controlled by the in-rush–limiting resistor (RH) that provides a very nonlinear *V*/*I*/*t* function. When cold, the resistance is very high, limiting the peak charging current of C2,3 when first turned on. Once the caps are charged, the resistance drops to a value conducive of the anticipated load current used. This eliminates the large spike of current through D1,3 that would possibly be a step transient that damages other components.

The rectified 340 Vdc is now fed to the half-bridge consisting of MOSFETs (Q2,3), where the voltage is chopped at a high frequency suitable for the primary windings of the small ferrite step-up transformer (T1). The input signal to Q1,2 is generated by the high-side driver (I2). It is important to note that the gate of Q2 is referenced at one-half the 340 Vdc. This is overcome by the ability of I2 to reference its ground by bootstrap capacitor C11, fooling Q2 into a mode where its gate only sees the correct drive voltage of 15 V by the holding ability of C11 maintaining a virtual ground. Q3 is referenced from the negative rail and operates normally. The power to I2 is via dropping resistors (R2AB) coming from the midsection of the voltage-doubling capacitors C2,3. Even though the driver chip has a built-in Zener diode, we choose a 1-W external one (Z1) as a safety.

Capacitor (C1,2) must be as close to the I2 as possible to supply the peak drive currents necessary for the gates of Q2,3.

Assembly

Assembly of the device is divided into four parts:

1. Control-board circuitry
2. Line-driver circuitry
3. High-voltage circuitry
4. Chassis/component fabrication and interconnecting wiring

1. Assembly: Control Board

Solder components and wires to a perf board, as shown in Figs. 1-3 and 1-4 (refer to the schematic of Fig. 1-2 for any needed clarification).

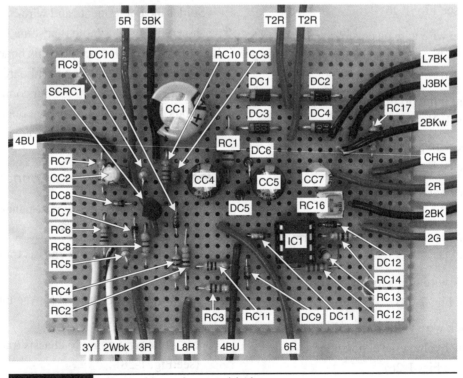

Figure 1-3 Control board, assembled.

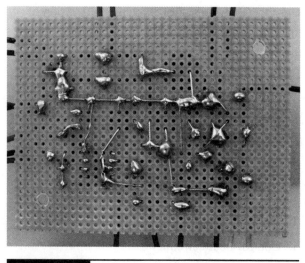

Figure 1-4   Control board wiring connections.

Control Board Wiring Connections

L8R: red wire going to line-driver printed circuit board (PCB) (see Fig. 1-7)

L7BK: black wire going to line-driver PCB (see Fig. 1-7)

3R: red wire going to front-panel stop switch S3 (see Fig. 1-26)

3Y: yellow wire going to front-panel stop switch S3 (see Fig. 1-26)

2Wbk: white wire with black stripe going to front-panel voltage-level potentiometer RC15/S2 (see Fig. 1-26)

5R: red wire going to front-panel stop switch S5 (see Fig. 1-26)

5BK: black wire going to front-panel stop switch S5 (see Fig. 1-26)

T2R: red wire going to chassis transformer T2 (see Fig. 1-28)

T2R: red wire going to chassis transformer T2 (see Fig. 1-28)

J3BK: black wire going to rear-panel 12-Vdc jack J3 (see Fig. 1-28)

2BKw: black wire with white stripe going to front-panel voltage-level potentiometer RC15/S2 (see Fig. 1-26)

CHG: green wire to chassis ground (see Figs. 1-27 and 1-28)

2R: red wire going to front-panel voltage-level potentiometer RC15/S2 (see Fig. 1-26)

2BK: black wire going to front-panel voltage-level potentiometer RC15/S2 (see Fig. 1-26)

2G: green wire going to front-panel voltage-level potentiometer RC15/S2 (see Fig. 1-26)

4BU: blue wire going to front-panel mode switch S4 (see Fig. 1-26)

4BU: blue wire going to front-panel mode switch S4 (see Fig. 1-26)

6R: red wire going to front-panel 12-V trigger switch S6 (see Fig. 1-26)

2. Assembly: Line Driver

The component layout is shown on the PCB in Fig. 1-5, with the wiring shown in Fig. 1-6 as an "x-ray" view (as if you're looking straight through the component side of the PCB). This can be built on a standard perf board, or the PCB is available for about $10 on our website: www.amazing1.com.

Solder the components and wires to the board as shown in Fig. 1-7 (note that some component locations are unfilled because this board is also used for other designs). If a perf board is being used, take a marker and trace out component locations on one side and wiring connections on the other before starting the soldering.

Line-Driver PCB Wiring Connections

1J: black wire going to front-panel power switch S1 (see Fig. 1-26)

2J: black wire going to chassis transformer T2 (see Fig. 1-28)

7J: white wire going to 115-Vac input power cord CO1 (see Fig. 1-26)

6J: white wire going to chassis transformer T2 (see Fig. 1-28)

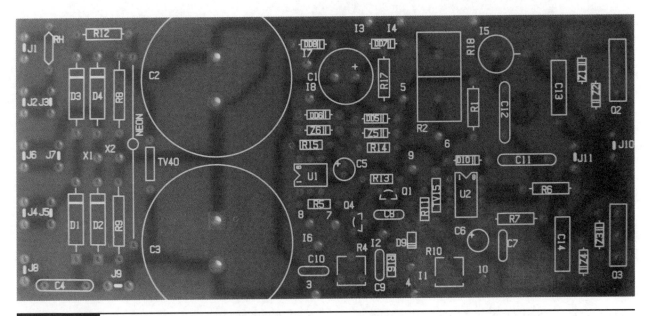

Figure 1-5   Line-driver PCB component layout.

8J: green wire going to 115-Vac input power cord CO1 (see Fig. 1-26)

CHASSIS GROUND: green wire going to chassis (see Figs. 1-27 and 1-28)

NEON: red and black wires going to front-panel neon light (see Fig. 1-26)

L8R: red wire going to control board (see Fig. 1-3)

L7BK: black wire going to control board (see Fig. 1-3)

1T: Litz wire going to transformer T1 (see Figs. 1-21 and 1-28)

1T: Litz wire going to transformer T1 (see Figs. 1-21 and 1-28)

Note that MOSFETs Q2 and Q3 are the most likely components to burn out, so instead of

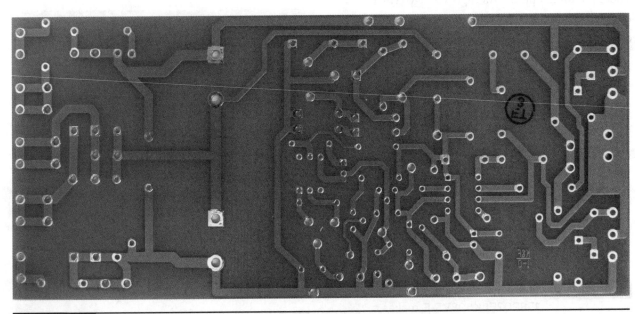

Figure 1-6   Line-Driver PCB wiring connections ("x-ray" view).

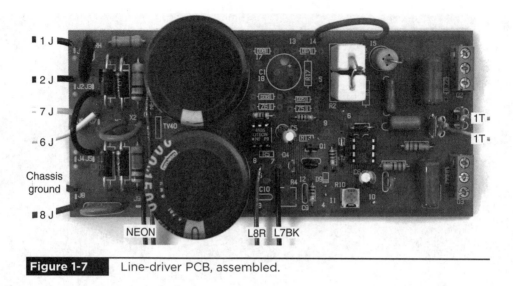

Figure 1-7	Line-driver PCB, assembled.

soldering these MOSFETs directly to the LINE DRIVER board, we instead solder a bracket with three screw-down terminals to hold the MOSFETs (Fig. 1-8).

The MOSFET terminals are then bent 90 degrees and secured to the aluminum chassis with screws (and thermal tape to increase heat dissipation) and can be easily replaced if they do happen to burn out (Fig. 1-9).

3. Assembly: High-Voltage Module

There are three options for high-voltage (HV) output modules: a *low-output range* that gives an output from 200 to 1000 V, a *medium-output range* that gives an output from 1000 to 5000 V, and a *high-output range* that gives an output from 5000 to 10,000 V.

The use of different HV output boards will require the adjustment of the potentiometer RC18 to match the output range.

3a. HV Module Assembly: 0.2-1kV Low-Output Range

Use this module for an output range of 200 to 1000 V (Fig. 1-10).

Figure 1-8	Mouser Part No. 158P02ELK508V3-E makes a good MOSFET socket.

Figure 1-9	MOSFETs in their brackets.

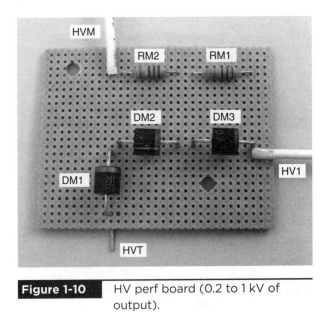

Figure 1-10   HV perf board (0.2 to 1 kV of output).

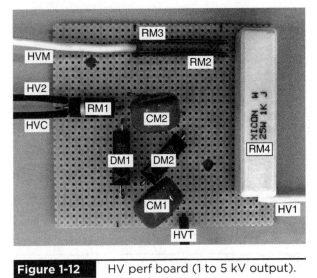

Figure 1-12   HV perf board (1 to 5 kV output).

HV Perf Board Wiring Connections (0.2 to 1 kV Low Output)

HV1: HV wire going to rear-panel red positive output jack HVJ1 (Fig. 1-27)

HVM: HV wire going to front-panel meter M1 (see Fig. 1-26)

HVT: HV wire going to transformer T2 (see Figs. 1-21 and 1-28)

Note that there is no wire from this module going to the HVJ2 jack (Fig. 1-27) on the rear panel; instead, the HVJ2 jack is wired to chassis

ground when using this low-output-range (200 to 1000 V) HV module (also see Fig. 1-2).

3b. HV Module Assembly: 1 to 5 kV Medium-Output Range

Use this module (Fig. 1-12) for an output range of 1000 to 5000 V

HV Perf Board Wiring Connections (1 to 5 kV Medium Output)

HV1: HV wire going to rear-panel red positive-output jack HVJ1 (see Fig. 1-27)

HVM: HV wire going to front-panel meter M1 (see Fig. 1-26)

HVT: HV wire going to transformer T2 (see Figs. 1-21 and 1-28)

HVC: HV wire going to chassis ground (see Figs. 1-27 and 1-28)

HV2: HV wire going to rear-panel black negative-output jack HVJ2 (see Fig. 1-27)

3c. HV Module Assembly: 5 to 10 kV High-Output Range

Use this module (Fig. 1-14) for an output range of 5000 to 10,000 V.

Figure 1-11   HV perf board (0.2 to 1 kV output) wiring.

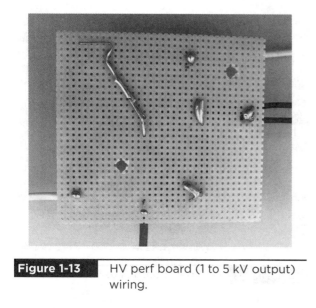

Figure 1-13 HV perf board (1 to 5 kV output) wiring.

HV Perf Board Wiring Connections (5 to 10 kV High Output)

The output wires are the same as those for the medium-output board in Fig. 1-12.

4. Assembly: Chassis and Components

4a. Chassis and Components Assembly: Transformer

The primary coil of the T1 transformer uses 80 turns of Litz wire for maximum efficiency;

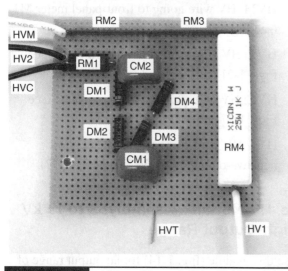

Figure 1-14 HV perf board (5 to 10 kV output).

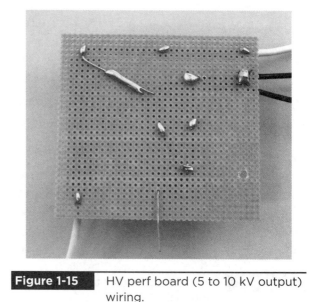

Figure 1-15 HV perf board (5 to 10 kV output) wiring.

however, heavy themaleze magnetic wire will work. To build this primary coil, lay the wire across a 1¾- × ¾-inch outside-diameter (OD) tube, with several inches of wire overhanging the edge (enough to reach the component to which the coil will be attached plus a little slack), and wrap the end of the tube with some tape to secure the wire in place (Fig. 1-16). If you don't have thin tape, just cut a strip to about 3/8-inch width, but it is important to use a thin strip of tape here because the bend of the wire should be close to the tube edge.

Bend the wire and start wrapping it around the tube (Fig. 1-17).

Keep the wire snug as it is wound (Fig. 1-18). Also keep track of the number of turns, either counting as the wire is wound or waiting until

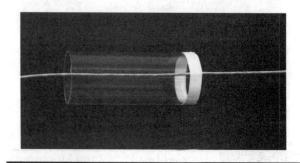

Figure 1-16 Start of transformer coil winding.

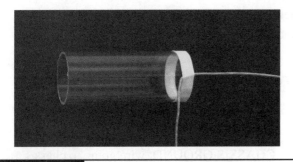

Figure 1-17   Bend wire and begin winding.

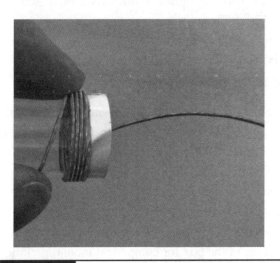

Figure 1-18   Keep winding snug.

the end is reached and then counting—whatever works best.

Once the wire gets to about 3/8 inch from the other end of the tube, wrap the entire length of the winding with two layers (for electrical insulation purposes) of Mylar tape to hold it in place (Fig. 1-19, step 3). Count the number of windings before it is wrapped because the wires are more difficult to count once they have been taped over. Then reverse direction (step 4), and continue winding back down the tube in the other direction (step 5; the wire still turns around the coil in the same orientation, but now it goes down the tube instead of up) until the full number of windings has been reached and then tape it all securely together so that the wire does not unwind (step 6). After the coil is finished with the correct number of total windings (80), it is good to wrap the entire coil once with Mylar tape to ensure that the windings stay tightly together in place and does not open up over time (see Fig. 1-20).

Next, cut two pieces of Mylar tape to act as a 2-mil air gap between the halves of the UU69 ferrite core (Fig. 1-20).

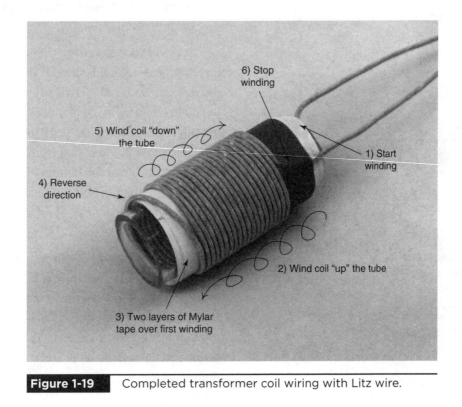

6) Stop winding

5) Wind coil "down" the tube

4) Reverse direction

3) Two layers of Mylar tape over first winding

1) Start winding

2) Wind coil "up" the tube

Figure 1-19   Completed transformer coil wiring with Litz wire.

Figure 1-20 Ferrite core with 2-mil Mylar taped air gaps.

The secondary coil uses 1000 turns of No. 32 magnetic wire, which is obviously a more involved fabrication. For this reason, we use a premanufactured and potted secondary coil (our Part No. COIL1000; see Fig. 1-21), but you also may wind this secondary coil if you prefer to fabricate it yourself. There are no special requirements (as with the primary) for insulating the layers of winding on this secondary coil—only to make 1000 turns.

With the primary and secondary fabricated, next, assemble the flyback transformer (Fig. 1-21). The top and bottom halves of the ferrite core can be

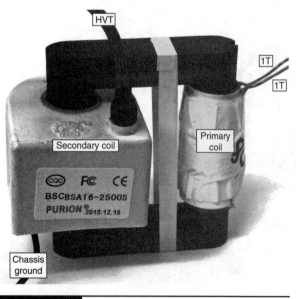

Figure 1-21 Assembled transformer.

held together with tape or a tie wrap (see Figs. 1-27 and 1-28).

Transformer T1 Wiring Connections

HVT: HV wire going to HV module (see Fig. 1-10, 1-12, or 1-14)

CHASSIS GROUND: wire going to chassis ground (see Figs. 1-27 and 1-28)

1T: Litz wire going to line-driver PCB (see Fig. 1-7)

4b. Chassis and Components Assembly: Chassis

The chassis can be fabricated from aluminum and the insulating sheet, undercarriage, and cover from sheet plastic. Alternately, a metal box (such as an electrical box) of appropriate dimensions may be substituted for the chassis (Fig. 1-22). But the plastic undercarriage and insulating plate will still need to be made, and holes will need to be cut into the face of the box for component mounting. Put rubber grommets in the holes of the aluminum chassis backplate through which wires will travel to protect them from fraying.

The plastic undercarriage (Fig. 1-23) serves three purposes: It holds the HV components at the rear, separate and electrically shielded from the main circuitry in the aluminum chassis; its bent-down edges provide "legs" to hold the box level and provide separation from a few bolts and wires that run along the bottom of the device; and it provides mounting holes to secure the cover in place.

Clear plastic is used for both the cover and the undercarriage to allow viewing of the neon safety lamp attached to the rear HV output jacks (upper-right section of the rear panel in Fig. 1-27). If an opaque cover is used, a hole may be drilled through the top cover such that the neon lamp can be pushed through the hole and still be seen to illuminate.

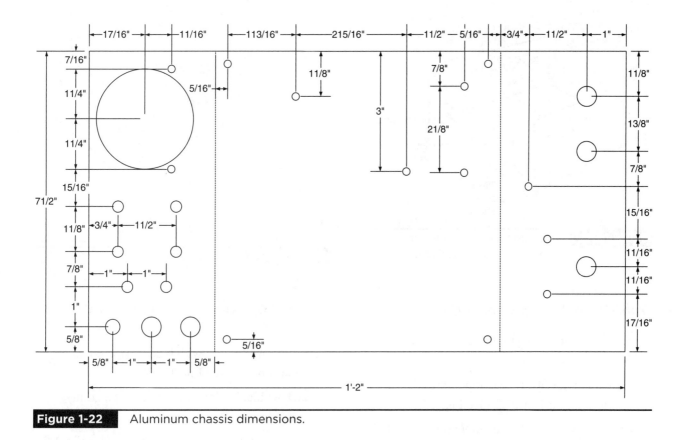

Figure 1-22   Aluminum chassis dimensions.

Then cut a plastic insulating sheet to fit inside the aluminum chassis—which should be approximately 7-3/8 × 7-3/8 in, but size according to the actual chassis dimensions (Fig. 1-24).

Then fabricate a clear plastic cover to go over the chassis and undercarriage, cut to fit (Fig. 1-25).

Next, attach components to the front panel, and wire them to their respective locations (Fig. 1-26).

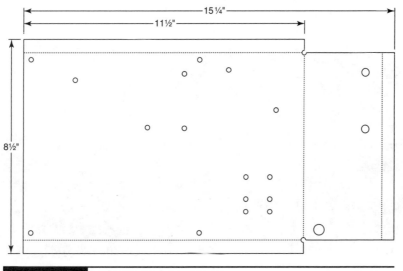

Figure 1-23   Plastic undercarriage.

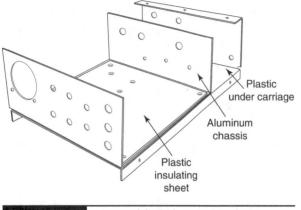

Figure 1-24   Assembled chassis, insulating sheet, and undercarriage.

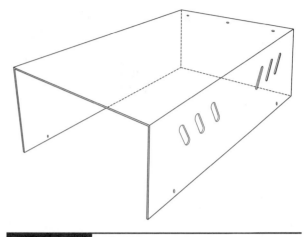

Figure 1-25   Plastic cover.

Front-Panel Wiring Connections

1J: black wire going to line-driver PCB (see Fig. 1-7)

7J: white wire of 115-Vac input power cord going to line-driver PCB (see Fig. 1-7)

8J: green wire of 115-Vac input power cord going to line-driver PCB (see Fig. 1-7)

NEON: red and black wires going to line-driver PCB (see Fig. 1-7)

3R: red wire going to control board (see Fig. 1-3)

3Y: yellow wire going to control board (see Fig. 1-3)

5R: red wire going to control board (see Fig. 1-3)

5BK: black wire going to control board (see Fig. 1-3)

2Wbk: white wire with black stripe going to control board (see Fig. 1-3)

2BKw: black wire with white stripe going to control board (see Fig. 1-3)

2R: red wire going to control board (see Fig. 1-3)

2BK: black wire going to control board (see Fig. 1-3)

2G: green wire going to control board (see Fig. 1-3)

4BU: blue wire going to control board (see Fig. 1-3)

4BU: blue wire going to control board (see Fig. 1-3)

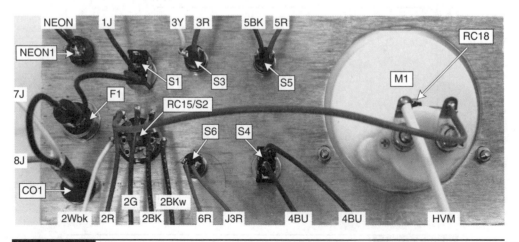

Figure 1-26   Front-panel components and wiring.

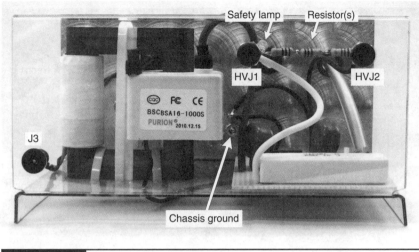

Figure 1-27 Rear-panel view.

6R: red wire going to control board (see Fig. 1-3)

J3R: red wire going to rear-panel 12-Vdc jack J3 (see Fig. 1-27)

HVM: HV wire going to HV perf board (see Fig. 1-10, 1-12, or 1-14 depending on which board is used)

Then attach components to the rear of the undercarriage (Fig. 1-27). If the low-output module is being used (see Fig. 1-2), solder one resistor RM3 in series with the safety lamp between the HV-positive jack HVJ1 and HV-negative jack HVJ2 (Fig. 1-27). If the medium- or high-power module is being used, solder three resistors RM5, RM6, and RM7 with the safety lamp between HVJ1 and HVJ2. Also note the 12-V output jack J3 (which is activated by the TRIGGER button on the front panel) in the lower left corner of the rear panel of Fig. 1-27.

With the boards and all chassis components secured in place, the charger is now ready to use (Fig. 1-28).

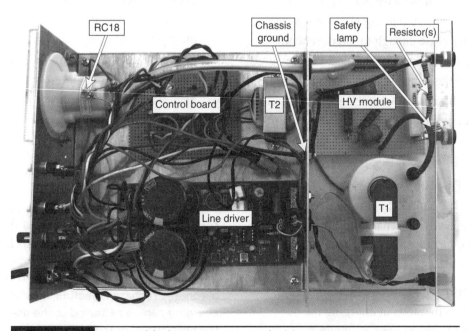

Figure 1-28 Assembled components and wiring.

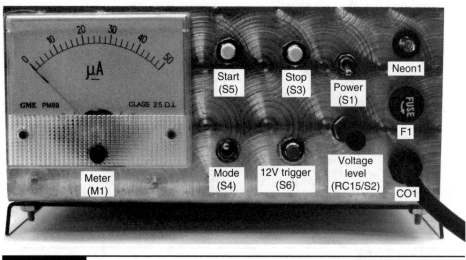

Figure 1-29   Front-panel controls.

Explanation of Controls

POWER (S1): Turns device on/off.

SAFETY LAMP: Not seen on the front panel, the *safety lamp* is attached to the rear HV output jacks instead (see Figs. 1-27 and 1-28) and indicates when the capacitor being charged exceeds 70 to 90 V (varies with temperature)— and the lamp will glow until the capacitor discharges below that voltage. [This lamp is placed at the rear because placing it at the front would require running HV wires across the circuitry close enough to interfere with the unit's operation. Therefore, a clear cover is used so as to see the lamp at the rear of the device. If you do not have (or want) a clear cover, a hole may be cut in the cover with the safety lamp poking through it so that it can still be seen to illuminate.]

NEON1: This lamp indicates that the unit is on and also will continue to glow until the circuit's internal capacitors discharge. *The low-voltage control circuitry remains on until the unit is unplugged.*

MODE (S4): The mode switch changes between two methods of charging the capacitor: Down is manual, and up is automatic. In the *manual (down)* position,

the capacitor is charged to the voltage set by the voltage-level dial (S2), and then the charger turns off. This setting is for larger capacitors that are scheduled to be discharged before much of the charge leaks off. In the *automatic (upper)* position, the capacitor is continually "topped off" to keep the voltage at the amount set by the voltage-level dial (S2)—the capacitor will discharge slightly and then automatically be recharged, repeating this cycle and keeping the capacitor charged up indefinitely. The mode switch only determines the *type* of charging (manual or automatic). Charging does not begin until the START button (S5) is pushed. (Note that factory-built negative-output units have the *automatic* mode deactivated.)

NOTE If the mode switch is left in the automatic position, the capacitor will immediately start to recharge after every full (or partial) discharge. This feature is for repetitive-firing modes but can be dangerous if the user forgets that the automatic-recharge mode is on. The STOP switch can be pushed to cease this automatic recharging action, at which point the charging can be restarted at the user's control for the next event, or the capacitor can be safely manually discharged.

START (S5): The START button must be pressed to start the charging action in either mode-switch setting. If the charging has been stopped, pressing this button will restart the charge. (This is convenient when charging a large storage capacitor and topping it off should it drop too far, for example.) The charging will still stop once the preset charge value is reached in the manual mode.

STOP (S3): The STOP button will stop the charging at any time, regardless of the mode setting. The START button must be pressed to resume charging.

> **NOTE** Reverify a positive earth ground. If in doubt, run a separate heavy-gauge No. 14 wire from the frame of the charger to a known earth ground. This is very important for safety. Never discharge the capacitor via the charger connection points. It is assumed that the user is aware of discharge ground loops and is experienced in high-energy technology.

12-V TRIGGER (S6): Pressing this button supplies 12 V to the rear 12-V jack and is used as a convenience switch to activate an external device (such as a trigger or ignitor).

VOLTAGE LEVEL (RC15/SC2): The voltage-control dial is used to set the charge voltage of the capacitor.

METER (M1): All meters read current; placing a proper resistor in series (as per the circuitry of this device) makes this meter draw its deflection current and thereby display voltage. This meter's units say "μA," but the circuitry has been modified such that it reads hundreds of volts. If desired, it is a simple matter to remove the two front screws, pull off the cover, and put a white sticker over the "μA" that says "dV" (for decavolt) or write "kV" and then white-out the zeros. Either way, this will give your meter accurate readings.

> **DANGER** Do not use this unit unless you fully understand high voltage and its hazards.

> **DANGER** A serious deadly shock hazard will exist when using with high-energy capacitors above 50 J.

Calculate *joules* by squaring the charge voltage and then multiplying by ½ the capacitance in microfarads and dividing by 1 million (to compensate for the measurement in microfarads). If the result is over 50 J, use extreme caution because improper contact can electrocute or cause serious burns.

$$J = \frac{cv^2}{2,000,000}$$

Operation Steps

First, select your capacitor, and use the Joulian formula above to calculate the maximum energy storage for determining if it will be hazardous when charged. Always verify that the capacitor is discharged before handling it. Then remove the safety shorting wire across the capacitor terminals before attaching it to the charger. (If you need to discharge, you can use an insulated screwdriver or a discharge resistor shorting wand for small electrolytic capacitors under 5 J. *(Larger capacitors require a properly made discharge probe depending on voltage and energy.)*

> **IMPORTANT** Verify that all controls are in their OFF position (POWER switch is turned off, VOLTAGE dial is turned off fully counterclockwise), and plug the device into a grounded wall jack. *The unit will prematurely start charging if the voltage-level control is not turned off. Place MODE switch in manual (down) position.*

1. Be sure that all controls of the charger are in their OFF position, and connect leads to capacitor, observing polarity (if any). Electrolytics are polarized. (If you require higher accuracy than the 3 percent front-panel meter, connect a proper range voltmeter across capacitor to monitor the charging voltage.)

NOTE The neon safety light across the output leads indicates voltages over 70 to 90 V (varies with temperature) and is placed inside a clear cover for maximum visibility.

2. Turn the charger on (POWER switch).

3. Rotate VOLTAGE LEVEL, and push and release the START button. The capacitor will gradually charge to the value set by the VOLTAGE LEVEL. It is recommended that you start at a very low level, such as turning the VOLTAGE LEVEL only past the first ON click and then pressing START to see how high the voltage rises. The STOP button may be pressed at any time to stop the charging (note that this will not discharge the capacitor but only stop any further charging). Otherwise, the voltage will continue to increase until it reaches the level set by the voltage-level potentiometer. (The voltage-level potentiometer cannot be "matched" to the voltage meter because each capacitor's response will differ according to its voltage and capacitance ratings.)

4. Repeat the process of increasing the VOLTAGE LEVEL and then pressing START button and watching the voltage increase on the front-panel meter until the required setting is reached (remembering that the STOP button can be pressed at any time to stop the charging). You may make notes on charge rate times for certain values of capacitance.

If the MODE switch is in the *manual* position, once the set charge is reached, the capacitor will slowly discharge and eventually totally discharge over a long enough period of time. However, if the MODE switch is flipped to the *automatic* position, then pressing START will keep the capacitor "topped off" at the level set by the voltage-level potentiometer. (And again, pressing the STOP button will stop the automatic charging, allowing the capacitor to slowly discharge over time.)

Do not allow the capacitor to charge beyond its volt rating, as indicated on the front-panel voltmeter. Obviously, larger values take longer charging time.

5. Always remember to verify that capacitors are fully discharged, disconnected from the charger, and have a shorting wire across their terminals when not being used. This is important if units are left exposed to contact or handling. Connect leads across the capacitor, and observe polarity, if any. Electrolytics are polarized.

6. There is no more data we can give as to the safe handling of charged capacitors because we do not know your specific application.

You are on your own and are assumed to understand the hazards of handling these very DANGEROUS and lethal amounts of electrical energy.

DANGER A serious deadly shock hazard will exist when using high-energy capacitors above 50 J.

Ref. No.	Quantity	Description
TABLE 1-1 Parts List		
		Line-Driver PCB
RH		In-rush limiter 47D15 3 amps hot
R2A/B		5-kΩ 5-W power resistor
R3,8		100-kΩ ¼-W film resistor (BR,BLK,YEL)
R6,7		15-Ω ¼-W film resistor (BR,GRN,BLK)
R10		5-kΩ horizontal trimpot
R11,14,15,16		1-kΩ ¼-W film resistor (BR,BLK,RED)
R12		470-kΩ ¼-W film resistor (YEL,PUR,YEL)
R18		10-Ω ¼-W film resistor (BR,BLK,BR)
C2,3		1500-µF 200-V vertical electrolytic
C4		0.01-µF 2-kV disk
C6,11		0.1-µF 50-V plastic
C7		10-µF 25-V vert electrolytic
C8,9		0.01-µF 50-V disk
C12		0.0015-µF 630-V polypropaline
C13,14		1.5-µF 250-V metalized film
D1-4		1N5408 3-amp 1-kV standard recovery diode
D10		1N4937 1-amp fast recovery diode
TV15		Z4744 15-V 1-W zener
Q1		PN2222 NPN TO92 general purpose
Q2,3		IRFP450 500-V *N*-type mosfet
U1		LM4N35 optoisolator
U2		IR2153 high-side driver
SOCK8X		8-pin connector for IR2153
SOCK6X		6-pin connector for LM4N35
SOCK247	2	3-pin connector mouser no. 158-P02ELK508V3-E for IRFP450S
		Control Perf Board
RC1		1-Ω ½-W film resistor (BR,BLK,GLD)
RC2,6,7		10-kΩ ¼-W film resistor (BR,BLK,OR)
RC3,4		1-kΩ ¼-W film resistor (BR,BLK,RED)
RC5,9		39-kΩ ¼-W film resistor (OR,WHI,OR)
RC8		470-Ω ¼-W film resistor (YEL,PUR,BR)
RC10		220-Ω ¼-W film resistor (RED,RED,BR)
RC11,17		10-Ω ¼-W film resistor (BR,BLK,BLK)
RC13,14		1-MΩ ¼-W film resistor (BR,BLK,GRN)
RC16		100-kΩ trimpot
RC18		2-kΩ trimpot
CC1		1000-µF 25-V vertical electrolytic
CC2		1-µF 25-V vertical electrolytic
CC3		0.1-µF 50-V plastic
CC4,5		100-µF 25-V vertical electrolytic
CC6,7		0.01-µF 50-V disk ceramic
DC1-4		1N4001 50-V 1-amp standard recovery diode
DC5,6		2 1N4735 6.2-V 1-W zener or 1 1N5242
DC7-10		1N914 general purpose silicon signal diode
DC11		Z5221 2.4-V ½-W zener

TABLE 1-1 Parts List (*Continued*)		
Ref. No.	**Quantity**	**Description**
		Control Perf Board
SCRC1		EC103A 100-V 0.8-amp SCR 200-mA gate
IC1		LM741 general purpose OP amp 8-pin dip
SOCK8X		8-pin connector for LM741
		200- to 1000-V High-Voltage Perf Board and Output
RM1,2		2.2-MΩ 1-W (RED,RED,GRN)
RM3		10-MΩ 1-W (BR,BLK,BLU)
DM1-3		FR607 3-amp 1000-V fast recovery diode
NEON2		Neon indicator lamp
J1		Red banana jack
J2		Black banana jack
		1000- to 5000-V High-Voltage Perf Board and Output
RM1		12-Ω 2-W (BR,RED,BLK)
RM2,3		50-MΩ 15-kV slim ox
RM4		1-kΩ 25-W power resistor
RM5,6,7		10-MΩ 1-W (BR,BLK,BLU)
CM1,2		0.001-µF 40-kV ceramic disk
DM1,2		UX-FOB 8kV 0.5-amp 0.5-µs high-voltage diode
NEON2		Neon indicator lamp
J1		Red banana jack
J2		Black banana jack
		10,000-V High-Voltage Perf Board and Output
RM1		12-Ω 2-W (BR,RED,BLK)
RM2,3		50-MΩ 15-kV slim ox
RM4		1-kΩ 25-W power resistor
RM5,6,7		10-MΩ 1-W (BR,BLK,BLU)
CM1,2		0.01-µF 40-kV ceramic disk
DM1-4		2CL2FM 2-kV 0.1-amp 100-nS high-voltage diode
NEON2		Neon indicator lamp
J1		Red banana jack
J2		Black banana jack
		Chassis-Mounted Parts
S1,S4		SPST mini toggle ¼-inch hole
RC15/S2		100-kΩ 17-mm pot/switch
S3,5,6		Push button N.O. momentary
T1		UL 60 core PRI 80-T no. 22 sec 1000-T no. 32
T2		12-V 0.3-amp 115-Vac small chassis mount
CO1		3-wire molded no. 18-gauge power cord
J3		Dc jack 2.5 mm
NEON1		Neon indicator lamp
M1		50-µA 2-inch panel meter
RC18		2-kΩ trimpot for M1 meter range adjust
FH1		5 × 20-mm panel fuse holder on panel
F1		3-amp fuse on panel

CHAPTER 2

Full-Feature Plasma Driver

Overview

High-performance driver is intended for the plasma-display manufacturer or experimental plasma scientist/experimenter. This is a reliable and flexible power supply to effectively drive light capacitive loads. Output power may be adjusted from 10 to 300 W. Utility unit operates from standard wall power (115 Vac, 60 Hz), measures $7 \times 4 \times 10$ inches, and weighs less than 5 lb. Intended for hydrogen production research, corona generation, energizing dielectric barrier junctions, ozone production, and other electrical and electrochemical applications.

Hazards

Moderate high voltage. Radiated plasma can cause preignition of flammables, nasty plasma burns, and damage to sensitive electronic equipment. Be careful of how you use test equipment. Eye protection should be worn when making, testing, and operating this device.

Difficulty

Requires intermediate skills in wiring and soldering, along with certain basic sheet-metal work to form and fabricate sheet-metal chassis, cover, and so on. Or you may be able to purchase a chassis/electrical box with close enough dimensions from electronics supply houses such as Mouser or Digi-Key, at which point all you need to do is drill the holes.

Electronic ability to use test equipment such as meters and a scope is necessary in completing this project. A good understanding of resonant circuitry and complex algebra would be very helpful.

Tools

Basic wiring, soldering, hand tools, bending apparatus, voltmeter, frequency meter capable of measuring from 10 to 150 kHz, and a *low-cost oscilloscope*.

You should familiarize yourself with Chap. 21 before building this circuit or any similar circuit without an isolation transformer. It advised to use the testing circuit jig as shown in Chap. 21 for this project because it verifies dangerous alternating-current (ac) grounds. Other amenities are powering a virgin circuit starting with low input voltage and slowly increasing, noting that any excessive current could be dangerous and totally wipe out your hard work!

Use

Provides an adjustable resonant output with independent power control. Produces energetic plasma of high frequency. Among other things, it has been used for research into nanoparticles and ionizing gases within a capacitive cell. This device has proved to be very popular with researchers in the fields of chemistry, medicine, energy, ozone, and water purification.

IMPORTANT   This is an advanced system intended for use by qualified personnel familiar with driving resonant circuits. Please make sure that you understand the operating parameters and use of this item before building it. However, you can repair any problems that should develop from misuses, being overdriven, and so on.

See "Duty Cycle Writeup" in Chap. 22 for the output characteristics and functionality of this power supply (Fig. 2-1).

- Independent voltage control from zero to maximum output of 20 kV peak-peak

- Independent current control from 5 percent to maximum output

- Independent frequency control from 20 to 70 kHz

- Large 3-inch ammeter monitors ac line current, indicates resonance peaking

- Adjustable waveform duty-cycle function allowing a wide range of operation

- Maximum output power over 300 W

- Short-circuit proof by built-in leakage reactance

- Measures $7 \times 4 \times 10$ inches and weighs less than 5 lb

Figure 2-1   Assembled plasma driver.

Assembly

Assembly is divided into four parts (the schematic is given in Fig. 2-2):

1. Line driver
2. Control board
3. Transformer
4. Chassis and interconnecting wiring

1. Assembly: Line-Driver Printed Circuit Board

The component layout is shown on the line-driver printed circuit board (PCB) in Fig. 2-3, with the wiring shown in Fig. 2-4 as an "x-ray" view (as if you're looking straight through the component side of the PCB). This can be built on a standard perf board, or the PCB is available for about $10 on our website: www.amazing1.com.

Solder the components and wires to the board as shown in Fig. 2-5 (note that some component locations are unfilled because this board is also used for other designs). If you are using a perf board, take a marker and trace out component locations on one side and wiring connections on the other before starting the soldering.

Line-Driver PCB Wiring Connections

S1J1: black wire going to front-panel main power switch S1

VA1J2: blue (or black) wire going to terminal C of the VA1 Voltage-Level Control main variac

VA1J7: white wire going to terminal A of the VA1 Voltage-Level Control main variac

PWR6: white wire going to PWR cord

S2X1: yellow wire going to front-panel Power Switch S2

T2J7: white (or black) wire going to T2 chassis transformer 115-Vac primary

S2X2: yellow wire going to front-panel Power Switch S2

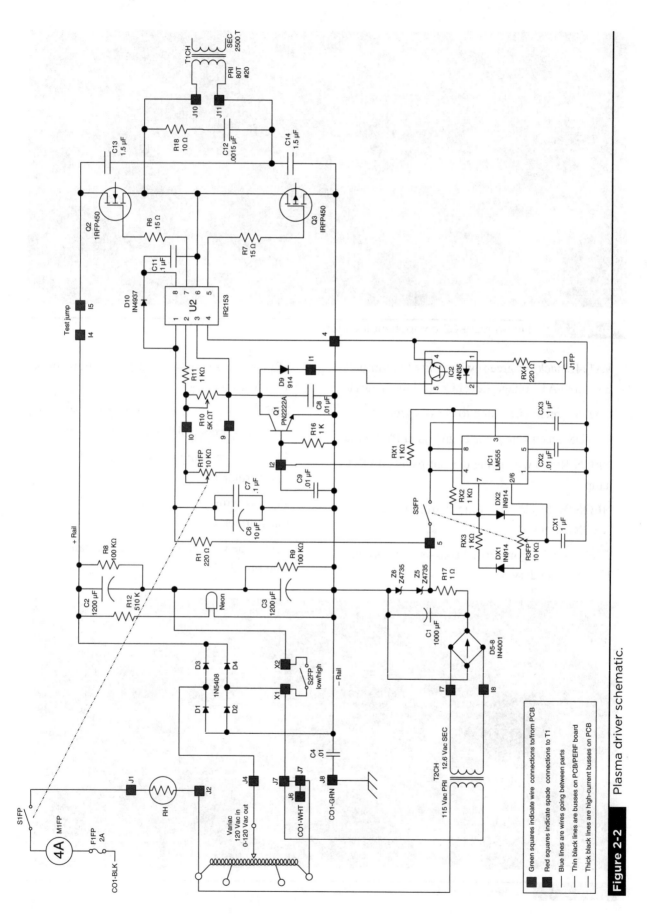

Figure 2-2 Plasma driver schematic.

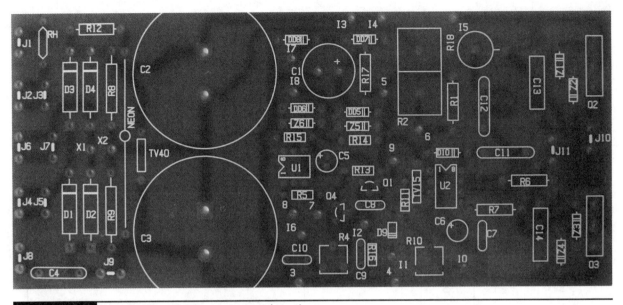

Figure 2-3 Line-driver PCB component locations.

VA1J4: black (or green) wire going to terminal E of the VA1 Voltage-Level Control main variac

GND8: green wire going to chassis ground

PGND8: green wire going to ground of PWR cord

NEON-R: red wire going to front-panel NEON1 lamp

NEON-B: black wire going to front-panel NEON1 lamp

T2i7: red wire going to T2 chassis transformer 12.6-V secondary

T2i8: red wire going to T2 chassis transformer 12.6-V secondary

RDC5: red wire going to front-panel RDC duty cycle/power control dial (see Fig. 2-20)

1T: Litz wire going to chassis flyback transformer T1

1T: Litz wire going to chassis flyback transformer T1

CB1: blue wire going to Control Board (see Fig. 2-6)

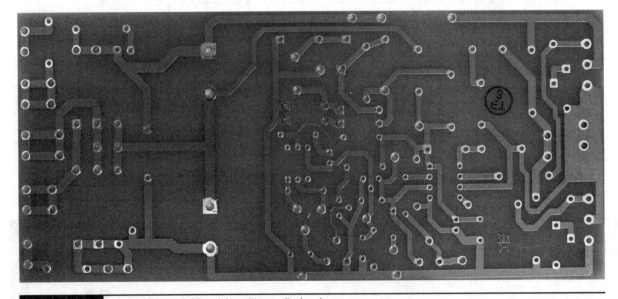

Figure 2-4 Line-driver PCB wiring ("x-ray" view).

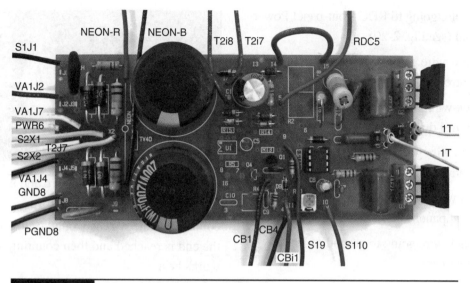

Figure 2-5   Line-driver PCB, assembled.

CB4: black wire going to Control Board
(see Fig. 2-6)

CBi1: red wire going to Control Board
(see Fig. 2-6)

S19: green wire going to front-panel S1
Frequency Dial (see Fig. 2-21)

S110: red wire going to front-panel S1 Frequency
Dial (see Fig. 2-21)

2. Assembly: Control Board

Build this (Fig. 2-6) from a small piece of standard
perf board (we do not currently use a PCB for this
control board), and drill a couple holes, as shown,
for later securing to the chassis (see Fig. 2-22).
See Fig. 2-20 for how the wires are soldered to the
control dial.

Control Board Wiring Connections
(Fig. 2-7)

CBi1: red wire going to line-driver PCB
(see Fig. 2-5)

CB1: blue wire going to line-driver PCB
(see Fig. 2-5)

CB4: black wire going to line-driver PCB
(see Fig. 2-5)

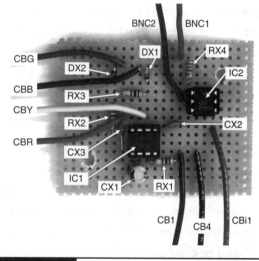

Figure 2-6   Control board, assembled.

Figure 2-7   Control board wiring.

CBR: red wire going to RDC front-panel Power Control dial (see Fig. 2-20)

CBY: yellow wire going to RDC front-panel Power Control dial (see Fig. 2-20)

CBB: black wire going to RDC front-panel Power Control dial (see Fig. 2-20)

CBG: green wire going to RDC front-panel Power Control dial (see Fig. 2-20)

BNC1: red wire going to center post of BNC jack on front panel

BNC2: black wire going to ground of BNC jack on front panel

3. Assembly: Transformer

The primary coil of the T1 transformer coil uses 80 turns of Litz wire for maximum efficiency, but heavy Themaleze magnetic wire will work. To build this primary coil, lay the wire across a 1¾- × ¾-inch outside-diameter (OD) tube, with several inches of wire overhanging the edge (enough length to reach the component to which the coil will be attached plus a little slack), and wrap the end of the tube with some tape to secure the wire in place (Fig. 2-8). If you don't have "thin" tape, just cut a strip to about 3/8-inch width, but it is important to use a thin strip of tape here because the bend of the wire should be close to the tube edge.

Bend the wire and start wrapping it around the tube (Fig. 2-9).

Keep the wire snug as it is wound (Fig. 2-10). Also keep track of the number of turns, either

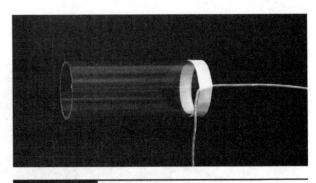

Figure 2-9   Bend the wire and begin winding.

counting as the wire is wound or waiting until the end is reached and then counting—whatever works best.

Once the wire gets to about 3/8 inch from the other end of the tube, wrap the entire length of the winding with two layers (for electrical insulation purposes) of Mylar tape to hold it in place (Fig. 2-11, step 3). Count the number of windings before it is wrapped because the wires are more difficult to count once they have been taped over. Then reverse direction (step 4) and continue winding back "down" the tube in the other direction (step 5; the wire still turns around the coil in the same orientation, but now it goes down the tube instead of up) until the full number of windings has been reached; then tape it all securely together so that the wire doesn't unwind (step 6). After the coil is finished with the correct

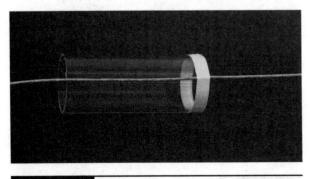

Figure 2-8   Start of transformer coil winding.

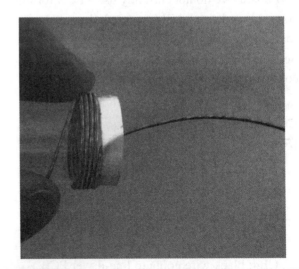

Figure 2-10   Keep the winding snug.

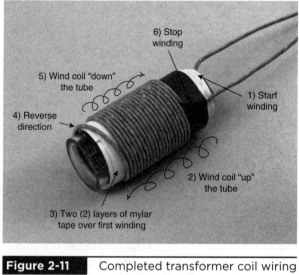

Figure 2-11 Completed transformer coil wiring with Litz wire.

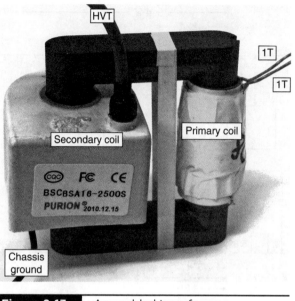

Figure 2-13 Assembled transformer.

number of total windings (80), it is good to wrap the entire coil once with Mylar tape to ensure that the windings stay tightly together in place and do not open up over time (see Fig. 1-20).

Next, cut two pieces of Mylar tape to act as 2-mil insulators between the halves of the ferrite core (Fig. 2-12).

The secondary coil uses 1000 turns of No. 32 magnetic wire, which is obviously a more-

involved fabrication. For this reason, we use a premanufactured and potted secondary coil (our Part No. COIL1000; see Fig. 1-21), but you also may wind this secondary coil if you prefer to fabricate it yourself. There are no special requirements (as with the primary) for insulating the layers of winding on this secondary coil, only to make 1000 turns.

With the primary and secondary fabricated, next, assemble the flyback transformer (Fig. 2-13). The top and bottom halves of the ferrite core can be held together with tape or a tie wrap. This transformer T1 is then secured to the back of the plastic undercarriage with a tie wrap (see Fig. 2-22).

Transformer T1 Wiring Connections

HVT: high-voltage (HV) output wire

CHASSIS GROUND: wire going to chassis ground (see Fig. 2-22)

1T: Litz wire going to line-driver PCB (Fig. 2-5)

4. Assembly: Chassis

Fabricate the inner chassis from a 1/16-inch-thick aluminum sheet and an insulating plate from

Figure 2-12 Ferrite core with insulating Mylar tape.

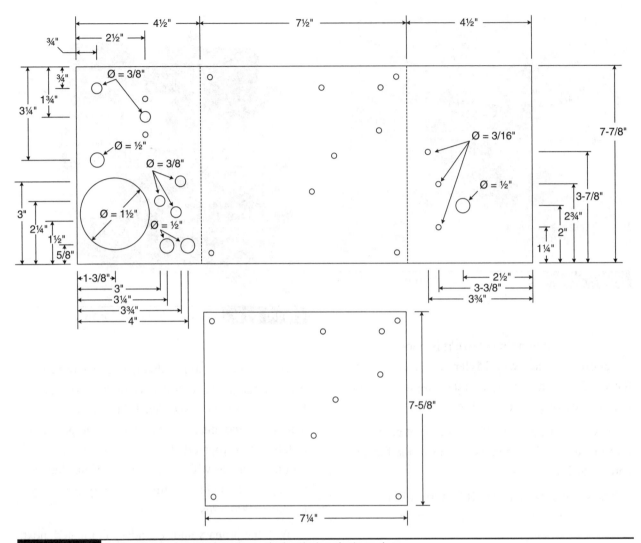

Figure 2-14   Aluminum inner chassis and plastic insulating plate.

1/16-inch plastic (Fig. 2-14). Just cut the sheets to size for now, but don't bend or drill them yet.

Then cut the undercarriage from a sheet of 1/16-inch plastic (Fig. 2-15); again, just cut the sheet to size for now and don't bend or drill anything yet.

With these three sheets cut (insulating sheet, aluminum chassis, and undercarriage), the easiest way to align the holes going through them is to stack and secure them together (Fig. 2-16) as they will be when the chassis is formed (Fig. 2-17). Again, don't drill or bend anything yet; just stack the sheets together as per Fig. 2-16 (using a marker to indicate the bend lines on the aluminum chassis will help with aligning the sheets). Once the sheets

are lined up properly, hold them together with a couple of wood clamps, and then the holes can be drilled through the three sheets and will line up perfectly.

The holes going through the insulating sheet don't need to be precisely located as long as they are positioned to secure the three sheets together (through the four corner holes) and hold the components (circuit board and transformers through the five inner holes) in place. Once these holes have been drilled through the three sheets, the remaining holes can be drilled through the aluminum chassis and plastic undercarriage and then bent to shape (Fig. 2-17). Then cut and bend

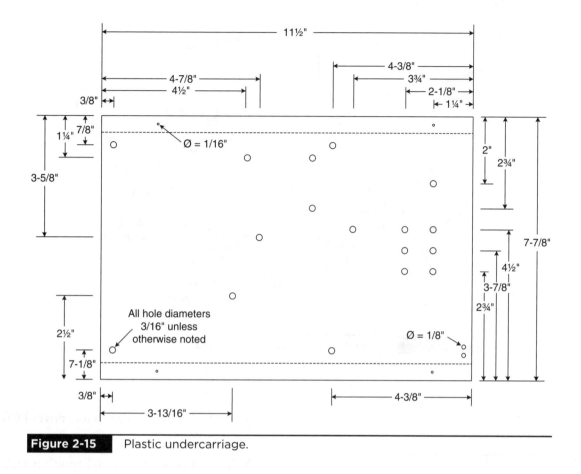

Figure 2-15   Plastic undercarriage.

the cover that will later go over the device and attach to the plastic undercarriage (Fig. 2-18).

Next, attach components to the aluminum chassis front panel and wire them to the circuitry and electronics (Fig. 2-19).

Front-Panel Wiring Connections

NEON-R: red wire going to line-driver PCB (see Fig. 2-5)

NEON-B: black wire going to line-driver PCB (see Fig. 2-5)

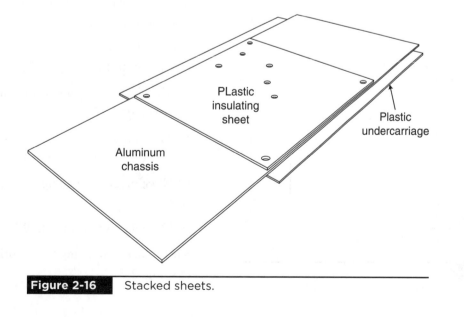

Figure 2-16   Stacked sheets.

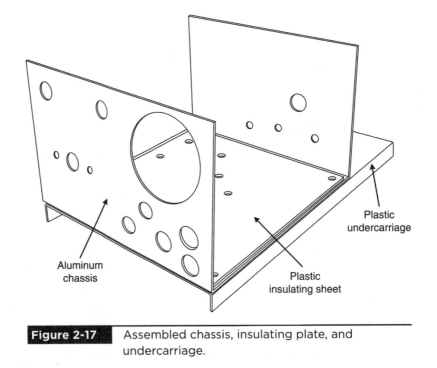

Figure 2-17 Assembled chassis, insulating plate, and undercarriage.

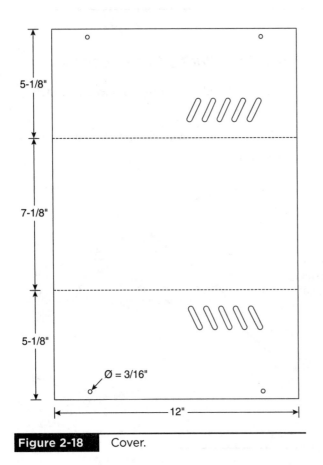

Figure 2-18 Cover.

S2X1: yellow wire going to line-driver PCB (see Fig. 2-5)

S2X2: yellow wire going to line-driver PCB (see Fig. 2-5)

VA1J7: white wire going to line-driver PCB (see Fig. 2-5)

VA1J2: blue (or black) wire going to line-driver PCB (see Fig. 2-5)

BNC1: red wire going from BNC center post to Control Board (see Fig. 2-6)

BNC2: black wire going from ground of BNC jack to Control Board (see Fig. 2-6)

T2BK: black wire going to chassis transformer T2 (see Fig. 2-22)

VA1J4: black (or green) wire going to line-driver PCB (see Fig. 2-5)

PGND8: green wire going to ground of PWR cord

PWR6: white wire going to PWR cord

S1J1: black wire going to line-driver PCB (see Fig. 2-5)

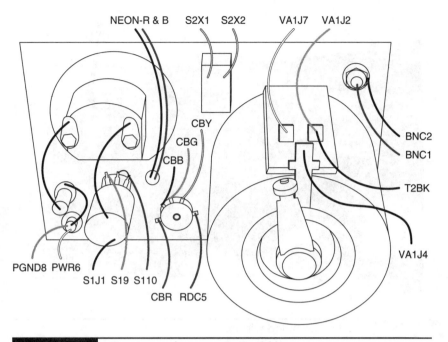

Figure 2-19   Front-panel component wiring.

S19: green wire going to line-driver PCB
(see Fig. 2-5)

S110: red wire going to line-driver PCB
(see Fig. 2-5)

CBR: red wire going to Control Board
(see Fig. 2-6)

CBB: black wire going to Control Board
(see Fig. 2-6)

CBG: green wire going to Control Board
(see Fig. 2-6)

CBY: yellow wire going to Control Board
(see Fig. 2-6)

RDC5: red wire going to line-driver PCB
(see Fig. 2-5)

Detail of the Power Control dial wiring is shown in
Fig. 2-20.

Detail of the Frequency Control dial is shown in
Fig. 2-21. Note the two terminals soldered together
where S110 connects.

With all the components in place and wired
together, this power supply is now ready for use
(Fig. 2-22). Note that the HVT is the high-voltage
output wire.

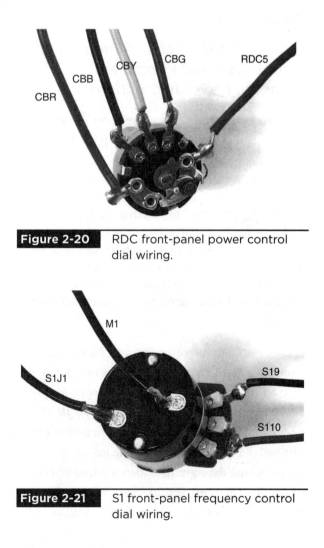

Figure 2-20   RDC front-panel power control
dial wiring.

Figure 2-21   S1 front-panel frequency control
dial wiring.

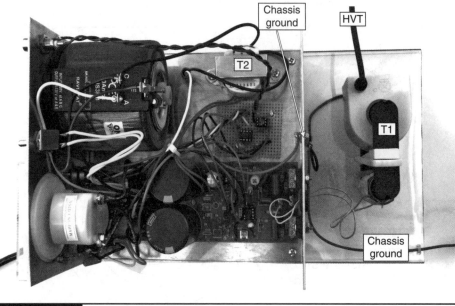

Figure 2-22   Fully assembled wiring.

Instructions

Hydrogen and Chemical Production, Corona Cell, and Plasma and Dielectric Driver Instructions

Intended for capacitive loads up to >1 μF and single-ended plasma gas displays up to 200 pF. For 115-Vac operation only. Use a 300-W step-down transformer for 220-V operation.

CAUTION   Running plasma displays and small objects can easily be damaged by the abrupt resonant power rise of this system. Always start with VA1 set at about 5 to 10 percent, and increase slowly. Use your own judgment on the damage point of the object used for load.

This useful high-frequency driver allows the user to tune to a capacitive load within the range of 5 pF to >1 μF. (*See data below on optional coils for higher values of capacitive loads up to >1 μF for other functions.*) This value is found in many corona cells and plasma-filled vessels. The maximum voltage across the capacitive load is a function of the circuit Q and can peak to levels that can destroy the output transformer and associated circuitry.

Please take note that the maximum voltage across a capacitive load is a function of the circuit Q and can peak to levels that can destroy the load under test, as well as the output transformer and associated circuitry. Therefore, the unit is not totally user-friendly and is intended for use by those experienced in powering up these resonant capacitive loads. Use caution because the output transformer can be easily damaged if allowed to spark over encapsulation.

Controls (Fig. 2-23)

VA1: voltage-level control

BNC: BNC jack for TTL modulation

S1/RFreq: Main power switch and frequency control

S2: HI/LO input-voltage switch; always start with this switch in the LO position.

Rdc: duty-cycle/power control

NEON1: power-on indicator and reset lamp

AMP: meter 0 to 5 amps for power input monitoring

FUSE: 3-amp "slow-blow"

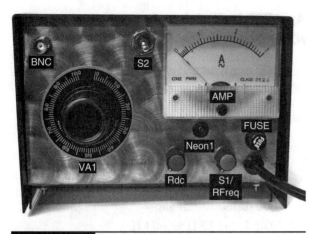

Figure 2-23 Front-panel controls.

Operation

1. Connect HV output lead to load. Note that output is referenced to chassis ground that is earth grounded via the green lead of the power cord.

2. Verify that the HI/LO switch is in the LO position, VA1 is full counterclockwise (FCCW), S1 is FCCW, and Rdc is fully FCCW/OFF.

3. Plug into a 115-Vac source, and rotate VA1 quarter range. Apply power via rotating the S1/RFreq control until it clicks on. Note that the NEON1 indicator lamp comes on (lamp comes on only when VA1 is set for above 70 to 90 Vac).

4. Slowly adjust S1/RFreq until the display starts to activate. *IMPORTANT:* This adjustment tunes the load capacitance to the unit's intrinsic inductance of the output transformer and should be set cautiously to a peak reading. It preferably should be set on the counterclockwise (CCW) side of the peak meter reading. Note that the S1/RFreq control increases frequency in the CCW direction. Now slowly rotate VA1 in a clockwise (CW) direction, noting the desired effect. Also note reading on the AMP meter for reference.

5. Repeat step 4 if necessary for required effect.

6. You may switch the HI/LO to HI for more power if the output is below 0.5 amp in the LO position. Do not exceed 3 amps, and check the transformer and circuit for heating and any excessive corona around the transformer or leads.

7. Now set Rdc to the desired current reading or display texture. Some loads may cause premature shutdown when Rdc is used.

Special Notes

Always check the output transformer for excessive heating, corona, or arcing (easiest to see in the dark). Do not allow it to operate in any of these states because the transformer will burn out. It may take up to 30 minutes for the transformer to overheat.

A SHUT DOWN circuit will trigger if output voltage becomes excessive, disabling the output. Reset will require power removal and waiting several minutes before reapplying, as indicated by the neon lamp fully extinguishing. This can be readjusted by slowly rotating the orange trimpot closest to the front-panel CCW for more output, or vice versa if the output is excessive. *Be careful in this adjustment because it can cause the transformer to burn out more easily.* You might want to contact the factory at tech@amazing1. com before moving this control from its factory setting.

Even though the output lead is rated for 40 kV, it must be clear of all conductive objects to prevent breakdown.

Certain loads may have different *Q* factors that will affect operation. *Q* factors are determined by the ratio of circuit reactance to resistance of the load. Reactance is the inductive and capacitive values at resonance. Resistance is determined by your load resistance, component losses, and the amount of useful corona or plasma ionization or whatever it is you need. *It might be wise to refresh your j operator or polar-notation math skills.*

Always attempt to operate radiofrequency (RF) just slightly below the current peak, as indicated on the AMP meter. This is especially important when operating above 1 amp to avoid overheating the switching transistors.

CAUTION Contact with the bare-metal controls and other objects may cause annoying burns. This is especially noticeable when powering single-ended plasma displays that are within several feet of the user. Insulated rubber tubing may be placed on the control shafts (Rdc and S1/RFreq) to help avoid these annoying shocks and burns.

Transformer Resonant Specifications for Those Who Wish to Use for Higher Load Capacity up to >1 μ

Transformers use our own standard tooled UU69 ferrite core with the following specs: CORE is $69 \times 39 \times 23$ mm, $u = 2000$, $Ae = 2.3$ cm^2, $Le = 22.9$.

Approximate Values for Load Capacity 2500-Turn Included Bobbin

Note that the primary air gap remains at 2 mils.

2500 turns, no gap = 16.5 H (6.2 H @ 60 kHz)	0.04 pF self- resonant/60 kHz
2500 turns, 4-mil gap = 8.5 H (3.2 H @ 60 kHz)	0.31 pF/60 kHz
2500 turns, 8-mil gap = 5.7 H (2.1 H @60 kHz)	0.46 pF/60 kHz
2500 turns, 20-mil gap = 2.9 H (1 H @ 60 kHz)	0.91 pF/60 kHz
2500 turns, no gap = 16.5 H (2.07 H @20 kHz)	3.8 pF/20 kHz
2500 turns, 4-mil gap = 8.5 H (1.06 H @ 20 kHz)	7.5 pF/20 kHz
2500 turns, 8-mil gap = 5.7 H (0.72 H @ 20 kHz)	11 pF/20 kHz
2500 turns, 20-mil gap = 2.9 H (0.36 H @ 20 kHz)	22 pF/20 kHz

Approximate Values for Load Capacity 1000-Turn Optional Bobbin

1000 turns, no gap = 2.6 H (+j.98 H @ 60 kHz)	2.7 pF/60 kHz
1000 turns, 4-mil gap = 1.36 H (+j.51 H @ 60 kHz)	5 pF/60 kHz
1000 turns, 8-mil gap = 0.9 H (+j.34 H @ 60 kHz)	7.8 pF/60 kHz
1000 turns, 20-mil gap = 0.46 H (+j.17 H @ 60 kHz)	15 pF/60 kHz
1000 turns, no gap = 2.6 H (+j.33 H @ 20 kHz)	24 pF/20 kHz
1000 turns, 4-mil gap = 1.36 H (+j.17 H @ 20 kHz)	46 pF/20 kHz
1000 turns, 8-mil gap = 0.9 H (+j.11 H @ 20 kHz)	70 pfd/20 kHz
1000 turns, 20-mil gap = 0.46 H (+j.05 H @ 20 kHz)	138 pF/20 kHz

The preceding possible combinations of the 1000-turn coil combined with the adjustable frequencies from 20 to 64 kHz allow you to resonate any capacitive cell from 2.7 to 138 pF and provide plenty of overlap just by changing the air gap in the secondary of the transformer.

Transformers with 1000 potted turns or a 2500-turn potted secondary coil will have a 2-mil air gap on each side. You may take apart and change the gap on the secondary side *only* to bring larger load capacities within tuning range. Leads must be as short as possible for low capacitive loads (<2.5 pF).

We have roughly calculated transformer secondary turns at the frequencies of 20 to 65 kHz for those who need to go to 1 nF, 10 nF, 100 nF, and >1 μF. The following list shows optionally available coils covering all ranges. The 2500-turn potted coil is supplied with all units. Hand-wound coils are easily pruned by removing or adding turns.

COIL2500S 2500-turn epoxy potted coil included on basic unit 20 to 65 kHz tunes up to 25 pF. Price: $39.50.

COIL1000S 1000-turn epoxy potted coil optionally available unit 20 to 65 kHz tunes 2.7 to 138 pF. Price: $39.50.

COIL560 560-turn nonpotted hand-wound coil tunes 100 to 1000 pF between 20 and 65 kHz. Price: $49.50.

COIL200 200-turn nonpotted hand-wound coil tunes 0.001 to 0.01 μF between 20 and 65 kHz. Price: $39.50.

COIL50 50-turn nonpotted hand-wound coil tunes 0.01 to 0.1 μF between 20 and 65 kHz. Price: $34.50.

COIL15 15-turn nonpotted hand-wound coil tunes 0.1 to >1 μF between 20 and 65 kHz, Price: $29.50.

TABLE 2-1 Parts List

Ref. No.	Quantity	Description	Order No.
			Parts for Line-Driver PCB
RH			In-rush limiter 47D15 3 amps hot
R1			220-Ω 1-W film resistor (RED, RED, BR)
R6,7	2		15-Ω ½-W film resistor (BR, GRN, BLK)
R8,9	2		100-kΩ 2-W film resistor (BR, BLK, YEL)
R10			10-kΩ horizontal trimpot
R11,16	2		1-kΩ ¼-W film resistor (BR, BLK, RED)
R17			1-Ω ½-W film resistor (BR, BLK, GLD)
R18			10-Ω 3-W mox noninductive (BR, BLK, BLK)
C1			1000-µF 25-V vertical-mount capacitor
C2,3	2		1200-µF 200-V vertical-mount capacitor
C4			0.01-µF 2-kV disk capacitor
C6			10-µF 25-V vertical-mount capacitor
C7			0.1-µF 50-V small blue polyester capacitor
C8			0.01-µF 100-V polyester
C9			0.01-µF 50-V ceramic disk
C11			0.1-µF 400-V vert met film capacitor
C12			0.0015-µF 630-V vert MET film capacitor
C13, 14	2		1.5-µF 250-V vert MET film capacitor
D1–4	4		IN5408 1-kV 3-amp SR diodes
D5–8	4		IN4001 50-V 1-amp rectifier
D9			IN914 60-V signal diode
D10			IN4937 1-kV 1-amp fast-recovery diode
Z5,6	2		Z4735 6.2-V zener diode
U2			8-pin dip IR2153 gate driver
Q2,3	2		IRFP450 power mosfet
Q4			PN2222A
SOCK247	2		3-pin connector mouser no. 158-P02ELK508V3-E for IRFP450
SOCK8X			8-pin IC socket for IR2153
			Parts for Control Perf Board
RX1-3	3		1-kΩ ¼-W film resistor (BR, BLK, RED)
RX4			220-Ω ¼-W film resistor (RED, RED, BR)
CX1			1-µF 25-V vertical-mount capacitor
CX2			0.01-µF 50-V ceramic disk
CX3			0.1-µF 50-V small blue polyester capacitor
DX1,2	2		IN914 60-V signal diode
IC1			8-pin dip timer LM555
IC2			6-pin dip OPTO ISO diode to NPN 4N35
SOCK6X			6-PIN IC socket for 4N35
SOCK8X			8-PIN IC socket for LM555

TABLE 2-1 Parts List (*Continued*)			
Ref. No.	Quantity	Description	Order No.
			Other Miscellaneous Parts
S1/R1FP			10-kΩ 24-mm pot/switch
S2			4A 125V switch
S3/R3FP			10-kΩ 17-mm pot/switch
T1CH			80T NO. 20 PRI 2500T SEC GP UU69 ferrite core
T2CH			115-V PRI 12.6-V sec
M1FP			3-amp AC meter
F1FP			2-amp fuse
CO1			3-wire cord
NEON			Neon indicator bulb
J1FP			Bnc connector
			Base
			Chassis
			Cover
			Neon bushing
			Cord clamp
			Screws
			Nuts
			Therma pads
			Insulator pad
			Double-sided tape
			Zip tie
			Wire bushing
			Wire nut

CHAPTER 3

Capacitor-Discharge Drilling Machine and Dielectric Tester

Overview

This high-voltage plasma device "drills" clean, round, and precise micron-sized holes.

Hazards

Nonlethal high voltage can produce a very painful electric shock. Eye protection must be worn because small pieces of material can be projected. Eye protection should be worn when making, testing, and operating this device.

Difficulty

Requires intermediate skills in wiring and soldering the PBKM charger, along with certain basic sheet-metal work to form and fabricate sheet-metal pieces and so on. You may be able to obtain these, with close enough dimensions, from electronics supply houses such as Mouser or Digi-Key, at which point all you need to do is drill the holes.

The driller requires 25 kV at low current to charge the system. You may build this charger or purchase it from our website: www.amazing1.com (Model PBK50).

Tools

Basic wiring, soldering, hand tools, bending apparatus, voltmeter capable of measuring to 40 kV. Electronic ability to use test equipment such as meters and a scope is necessary in completing this project.

Use

Use as a demonstration, science, or specialty project, providing a low-cost method of making small, round, and clear pin-sized holes in most nonconductive materials. Start by using an old paper business card. These small holes are excellent for optical projects or just research and development (R&D). Experiment with different materials and thicknesses, different sizes, materials of discharge electrodes, voltage values, and so on.

This useful device is intended for anyone requiring clean, micron-sized holes in nonmetallic materials. This unit will drill through a 1/8-inch-thick stack of business cards in one shot. It is great for anyone doing optical and laser studies requiring micron-sized holes. It is also great for experimenters validating the dielectric and insulating integrity of certain materials.

Figure 3-1 shows the device without protective shielding of high-voltage parts, but a simply made cover can be added if desired. Improper contact can produce a very painful shock but will not burn or cause injury. Plasma energy is 20 kV at 1/5 J.

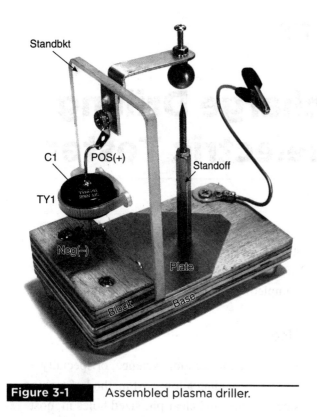

Figure 3-1 Assembled plasma driller.

Assembly

1. Fabricate the parts as shown in Fig. 3-2. The dimensions as shown are for the parts we used.

2. Position the PLATE onto the BASE. Next, position the BLOCK on top of the BASE, as shown in Fig. 3-1, and using two No. 6 × ¾-inch wood screws, assemble the pieces together. Note that the screw with the ground lug LUG6 for lead from C1 must make a good electrical contact with the PLATE when going into hole H—this is best accomplished by drilling the hole H in the aluminum PLATE at 3/32-inch diameter, which is smaller than the No. 6 screw going into it such that there will be a snug connection between the sheet-metal screw and the PLATE.

3. Attach the STANDOFF via an SW634 × ¾-inch screw from the underside of the BASE.

4. Attach the STANDBRACKET via SW612 No. 6 × ½-inch wood screws into the side of the BLOCK.

5. Fasten capacitor C1 with a 6-inch nylon tie wrap TY1.

6. Attach the aluminum BRACKET via two 6-32 × ½-inch screws and nuts. Note that the bottom screw has LUG6 for connection to C1. The LUG6 lugs are also used for connecting the wire leads from the charger PBK50 (below).

7. Attach BALL.5 to the BRACKET via a 6-32 × 1-inch screw and a top and bottom jam nut.

8. Sharpen the end of a 6-32 × 1½-inch screw to a symmetrical conical point. Remove the head and reshape the threads to screw into the STANDOFF as shown in Fig. 3-1. Note the jam nut securing the position. Adjust the spacing between BALL.5 and the reworked screw to ¼ inch. This is the firing gap and may be readjusted when you become more familiar with operation.

9. Assemble the sample holder using an alligator clip attached to a piece of No. 12-14 solid wire, and secure it to the base section using a large ¼-inch-hole solder lug No. LUG1420. This arrangement provides a flexible method of positioning the material to be drilled with your hands away from the high-voltage points. You may dispense with this step but will want to use caution to avoid a very painful (yet nonlethal) shock.

10. Finally, attach the rubber feet #FEET.

Operation

1. Adjust the drill gap spacing to ¼ inch. This is a good start, and it may be adjusted more as long as capacitor voltage rating is not exceeded.

2. Connect up a high-voltage current-limited power supply across the capacitor, as shown by the plus (+) and minus (−). You should use our PBK50, as shown in this chapter. You will notice that the output is through alligator-clip leads, with a high-voltage lead for the positive connection and a normal wire for the ground connection. This provides flexibility so that

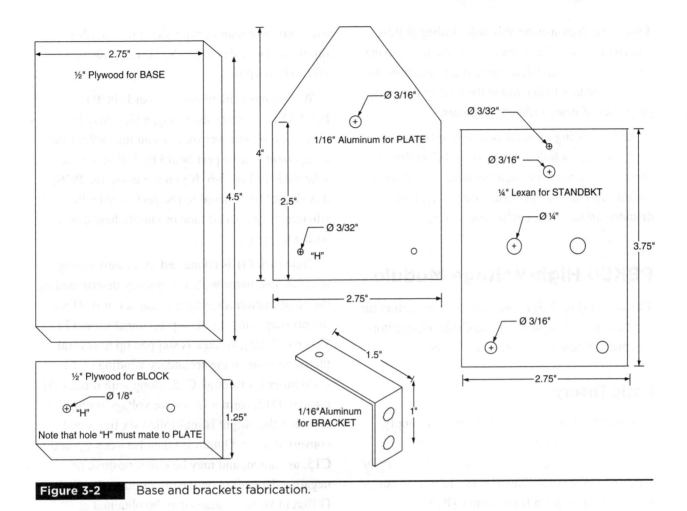

Figure 3-2   Base and brackets fabrication.

the unit can be used in other applications. You also may use banana jacks and plugs for a more permanent installation.

3. The direct-current (dc) jack on the PBK50 allows operation with a wall adapter or connection to a 12-Vdc battery or other 12-Vdc source. Our Model No. 12DC.3A is supplied with the PBK50.

4. The PBK50 is an unregulated source and will keep adding voltage to the capacitor as long as the push-button switch is held on or the drill gap fires. This is important to monitor because the capacitor could overcharge and short out if the drill gap voltage exceeds its voltage rating.

Another thing to watch is that the material being drilled by the capacitor-discharge machine should be paper, cardboard, wood, or a similar material.

If plastics are used, they must be thin enough to allow breakdown or the voltage could exceed the capacitor rating.

For more precise applications, you may consider using our optional HV350R regulated with a panel voltmeter. This will allow you to preset the voltage at any amount up to 35 kV. You also may consider using a larger capacitor, such as one of those listed in the parts list.

Applications

This device can create reasonably round holes in the diameter range of tens of microns (1 μm = 0.03937 mil). These hole sizes can be used in optical devices and other scientific applications. This is an excellent device that demonstrates this hole-drilling ability, and you can vary the hole sizes and shapes by using larger capacitors and different gap settings. You also may use the unit to determine the relative dielectric properties of many different materials.

An interesting exercise is to take a business card and blast a hole in it. Next, stabilize the card, shine a green laser through the hole, and observe the interference bands. Experiment with how distance affects the patterns and so on.

PBK50 High-Voltage Module

This device (Fig. 3-3) is used to charge capacitors (in this case, the C1 capacitor of the CDR10 Capacitor-Discharge Machine and Dielectric Tester).

Basic Theory

A miniature high-voltage (HV) power supply produces approximately 15 kV at several hundred microamps from a 7- to 12-V battery or other power source. This HV module uses our inexpensive step-up transformer (Part No. 28K077), which may be purchased from our website (www.amazing1.com), or you may

use your own transformer that takes a 12-V input and provides 2000-V, 10-mA, 20- to 100-kHz output.

We use our printed circuit board (PCB) PCPFS5 to simplify assembly, which may be purchased on our website, or you may solder the components into a perf board by following the schematic in Fig. 3-6. If you are using the PCB, this should be adhered to the perf board with silicone or two-sided tape or simply held down with a tie wrap.

Transistor Q1 is connected as a free-running resonant oscillator with a frequency determined by the combination resonance of capacitor (C3) and the primary winding of step-up transformer (T1). This oscillating voltage is stepped up to several thousand volts in the secondary winding of T1. Capacitors C4 through C15, along with diodes D1 through D12, form a full-wave voltage multiplier in which the output is multiplied six times and converted to dc. Output is taken between C5 and C15, as shown, and may be either positive or negative depending on the direction of the diodes. Different voltage values may be obtained at various taps of the capacitors. Note the schematic showing connections for taps to the image tube.

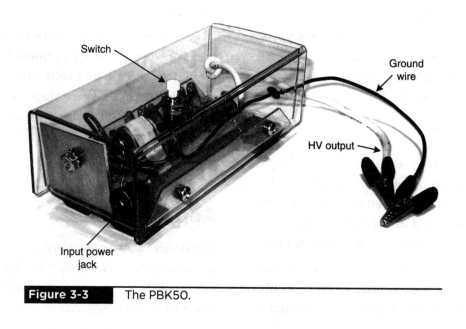

Switch

Ground wire

HV output

Input power jack

Figure 3-3   The PBK50.

The base of Q1 is connected to a feedback winding of T1, where the oscillator voltage is of proper value to sustain oscillation. Resistor (R2) biases the base into conduction for initial turn-on. Resistor (R1) limits the base current, whereas capacitor C2 speeds up the turning off of Q1 by supplying negative bias. Capacitor (C1) bypasses any high-frequency energy. Input power is supplied through switch (S1) via a snap-in battery clip.

PBK50 Assembly

The enclosure to hold the PBK50 can be built from two pieces of bent plastic (Fig. 3-4) or by slipping the circuitry into a plastic tube that is at least 1.5 inches in inner diameter (ID) and 7 inches long (or whatever accommodates your perf board) with a couple caps on the ends of

the tube, the power jack on one side, and the HV leads on the other.

The schematics and wiring are shown in Figs. 3-6 through 3-9.

Circuit Assembly

1. Lay out and identify all parts and pieces; check the parts list. Note that some parts sometimes may vary in value. This is acceptable because all components are 10 to 20 percent tolerance unless otherwise noted. A length of buss wire is used for long circuit runs.

2. Fabricate the perforated circuit board as shown in Fig. 3-8. Enlarge the holes as follows: thirteen 1/16-inch holes at the junction of the diodes and capacitors in the multiplier (Fig. 3-9) and seven 1/8-inch holes for mounting switch S1 and external connection leads.

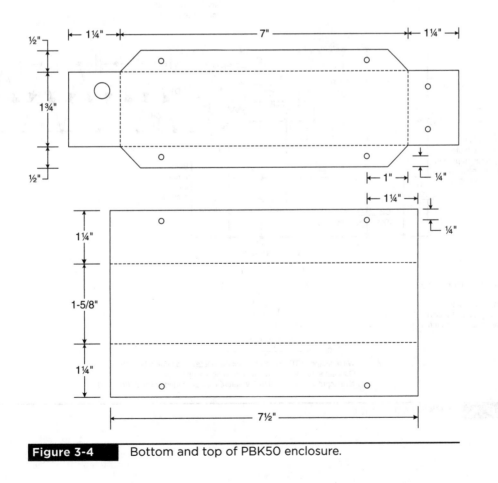

Figure 3-4   Bottom and top of PBK50 enclosure.

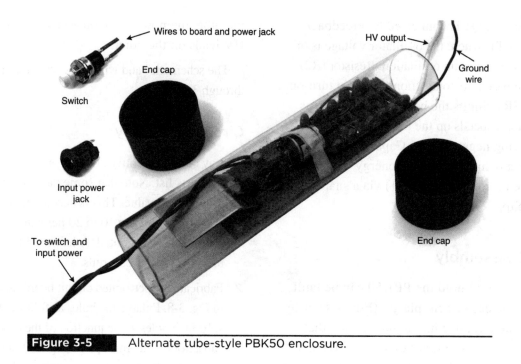

Figure 3-5 Alternate tube-style PBK50 enclosure.

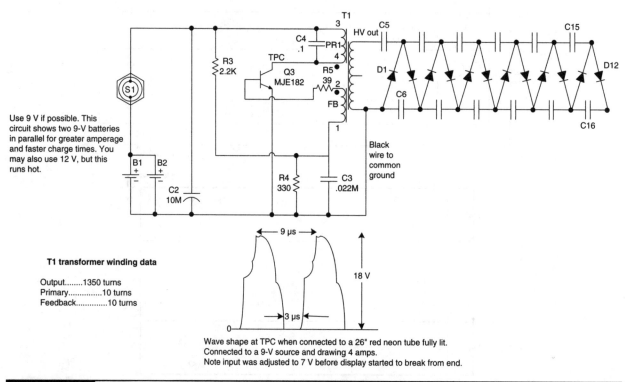

T1 transformer winding data

Output........1350 turns
Primary...............10 turns
Feedback..............10 turns

Wave shape at TPC when connected to a 26" red neon tube fully lit.
Connected to a 9-V source and drawing 4 amps.
Note input was adjusted to 7 V before display started to break from end.

Figure 3-6 PBK50 schematic.

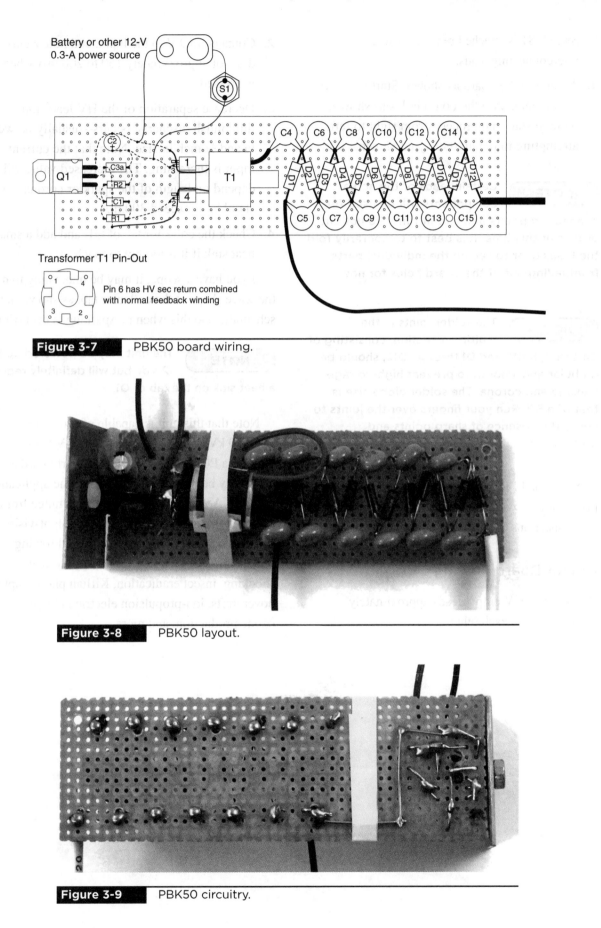

Figure 3-7 PBK50 board wiring.

Transformer T1 Pin-Out

Pin 6 has HV sec return combined with normal feedback winding

Figure 3-8 PBK50 layout.

Figure 3-9 PBK50 circuitry.

3. Switch S1 is attached remotely using interconnecting leads.

4. Assemble the board as shown. Start to insert components into the board holes as shown. Note to start and proceed from right to left, attempting to obtain the layout shown.

NOTE Certain leads of the actual components will be used for connecting points and circuit runs. Do not cut or trim at this time. It is best to temporarily fold the leads over to secure the individual parts from falling out of the board holes for now.

NOTE The solder joints in the multiplier section, consisting of C4 through C15 and D1 through D12, should be globular and smooth to prevent high-voltage leakage and corona. The solder globe size is that of a BB. Run your fingers over the joints to verify the absence of sharp points and protrusions.

Note that T1 is lying on its side and uses short pieces of buss wire soldered to its pins as extensions for connections to the circuit board.

Circuit Board Testing

1. Separate HV output leads approximately 1 inch from each other.

2. Connect up 9 V to the input, and note a current draw of approximately 150 to 200 mA when S1 is pressed.

3. Decrease separation of the HV leads until a thin bluish discharge occurs—usually between ½ and ¾ inch in size. Note that the current input is increasing. The increased value will depend on length of the spark or corona but should not exceed 300 mA.

4. Check the collector tab of Q1, and add a small heat sink if it is too hot to touch.

If you have a scope, it may be interesting to note the wave shape at the collector tab, as shown in the schematic. Do this when no sparking is occurring.

NOTE The unit may be powered up to 12 Vdc but will definitely require a heat sink on the tab of Q1.

Note that this unit is capable of producing 10 to 20 kV from a small standard 9-V battery. It is built on a PCB or a small piece of perf board and can easily be housed or enclosed as the application requires. Applications for this device range from powering image-converter tubes for night-vision devices to ignition circuits for flame-throwing units, capacitor charging for energy storage, shocking, insect eradication, Kirlian photography, hovercrafts, ion-propulsion electric field generators, ozone production, and more.

Info Part	Quantity	Vend No.	Description	Part #
TABLE 3-1 Parts List for CDR10 Capacitor-Discharge Drilling Machine*				
Basefab		Infofab	3/8-inch plywood, as shown in Fig. 3-2	
Blockfab		Infofab	3/8-inch plywood, as shown in Fig. 3-2	
Plateefab		Infofab	1/16-inch aluminum, as shown in Fig. 3-2	
Standbktfab		Infofab	¼-inch polycarbonate piece	
Bracketfab		Infofab	0.063-inch aluminum	
Standoff632175		Surplus	6-32 × 1¾-inch metal standoff for drill	
SW6-32X1.5	2	Infofab/ Cstores	6-32 × 1½-inch screw fabbed to point for GRD electrode drill	
SW6-32X.5	2		6-32 × ½-inch PH phil and nut	
SW612	2		No. 6 × ½-inch wood or sheet-metal screw	
SW675	2		No. 6 × ¾-inch wood or sheet-metal screw	
Nut6	5		No. 6 KEP nuts	
Ball.5			½-inch brass ball 6-32 hole	
Feet	4		Stick-on rubber feet	
Lug6	2		No. 6 solder lug	
Cap1			0.001 mF/20 kV shown or use 30 kV for higher-voltage discharges	0.001-mF/20-kV DKA 0.001-mF/ 30-kV DKA
Lug1420			¼-inch 20 large ring lug	
Tye6			6-inch nylon tie wrap to secure C1 to Standbktfab	
WR14	6		No. 14-12 solid buss wire	
Allclipmed			Medium-sized alligator clip	
SW-3X375			No. 8 × 3/8-inch sheet-metal screw	
Pbk5		Optional	Recommended charger for cdr1 with 12-Vdc/0.3 amp adapter	#PBK50
HV350R		Optional	For more precise and higher-energy applications	#HV350R
0.001-µF/40-kV DKT-2		Optional	0.001 µF at 40 kV for four times energy discharge than 20 kV discharge energy 0.8 J	0.001-µF/40-kV DKT-2

*Most parts should be available through electronics or hardware stores, but those more difficult to acquire are listed with a "Part #" and are available through www.amazing1.com if needed.

TABLE 3-2 Parts List for PBK50 Capacitor Charger		
Part	**Quantity**	**Description**
		Voltage Multiplier
C5–16	12	200–270-pF/3-kV plastic disk capacitor
D1–12	12	6-kV 100-NS HV avalanche diode
S1		Push-button switch
PB1		8- × 1½-inch perf board with 0.1- × 0.1-inch grid
CL1	2	Snap-on battery clip
WR22		24-inch length of no. 22 vinyl hook-up wire
WRHV20		12-inch 20-kv silicone wire
		PFS Printed Circuit Board
R3		2.2-kΩ ¼-W resistor (RED, RED, RED)
R4		33-Ω ¼-W resistor (OR, OR, BR)
R5		39-Ω ¼-W resistor (OR, WH, BLK)
C2		10-μF 25-V electrolytic capacitor
C3		0.022-μF 250-V plastic capacitor
C4		0.1-μF 400-V metal polypropylene capacitor
Q3		MJE182 power tab *npn* transistor
T1		Ferrite high-voltage transformer 28K077
CL1,2 2		Battery snap connectors
PCPFS5		Printed circuit board PCPFS5
HS1/SW1		Heat-sink bracket and no. 6 × ¼-inch nylon screw
Busswire		3-inch piece of buss wire for connecting pins of T1
		PBK50 Enclosure
Plastic 1		10- × 7-inch plastic/lexan sheet 1/16 inch thick
or		
tube		8-inch-long × 1½-inch-diameter plastic tube
cap 2		1-5/8-inch end cap to enclose tube CAP158

CHAPTER 4

Capacitor Exploder

Overview

A wire and part exploder, magnetizer, object accelerator, and high-power magnetic pulse generator. This device allows you to determine the explosion point of many components when operated in an overvoltage fault condition. This testing and evaluation is very important for components used as suppressors and absorbers, and for many other protective electrical components. Thin aluminum wire actually will detonate and vaporize!

Also, many ferrous objects may be magnetized by winding a coil with the proper number of turns and wire size over the target object and switching the discharge through this coil.

A very interesting electromagnetic reaction can be created as a demonstration of the Lorentz force. You can easily build a kinetic gun that shoots objects at a moderate velocity using this reaction.

Hazards

Dangerous high voltage must be handled carefully. Eye protection should be worn when making, testing, and operating this device because small pieces of material can be projected at high speeds, even though the device is shown built with an explosion shield. Work on nonconductive floors such as wood or concrete. Always use the "one hand in the pocket rule" when handling high-voltage (HV) devices.

Difficulty

Requires intermediate skills in wiring and soldering, along with certain basic sheet-metal work to form and fabricate sheet-metal parts, plastic cover, and so on. Most of these materials can be obtained at your local hardware store. Electronic parts can be obtained from electronics supply houses such as Mouser or Digi-Key or on our website (www.amazing1.com).

This capacitor exploder requires 450 V to charge the system. You may build a charger as shown in this chapter or purchase one from our website (www.amazing1.com, Model CHARGE800).

Tools

Basic wiring, soldering, hand tools, bending apparatus, and voltmeter capable of measuring to 1000 V. Electronic ability to use test equipment such as meters and a scope is necessary in completing this project.

Use

Provides a low-cost method of experimenting with various electrical parts, pieces of thin wire of different materials, or other objects and observing how they explode.

DANGER A serious deadly shock hazard will exist when using with high-energy capacitors above 50 J. Safety glasses are mandatory.

The unit uses a 6300-μF, 400-V polarized electrolytic capacitor. The project contains both electric and explosion shields, but this is still a very dangerous device that requires HV handling skills. The energy storage is 360 J at 400 V *and has over six times the energy necessary to cause a fatal shock if handled improperly.*

The charging circuit CHARGE800V must be powered by ungrounded batteries. This eliminates possibly dangerous ground currents because the electrical system is *ungrounded.* Although this does indeed prevent dangerous ground currents, please note that the system still can cause severe shocks and burns. When the neon safety lamp (Fig. 4-1) is glowing, this indicates that there is 70 to 90 V still on the capacitor, and caution must be used.

The storage capacitor C1 is charged by an ungrounded battery-powered inverter such as our

CHARGE800V being powered by 8 AA cells or a larger 12-V battery. The stored capacitive energy is switched into the target load via a mechanical contactor tripped by the removal of an insulating TRIGARM of plastic. This piece is placed between the spring-loaded contacts before the charging period. It is then quickly removed, causing the contacts to snap together, now dumping all the energy from the capacitor into the target load in a very short period of time. This action causes a high amount of peak power to the load.

Assembly

NOTE   Use the figures as a general guide. Even though our drawings are shown with dimensions, it is a good idea to verify the fitting before final fabrication of the actual part and based on the size of your capacitor.

The dimensions shown are used to build our units, but we suggest that you trial fit parts and drill holes as you go. Study the figures for parts placement, and verify that all fabricated parts match. Be sure to make the protective blast- and shock-protection shields.

1. Fabricate the explosion shield as shown on Fig. 4-2 from 1/16-inch clear polycarbonate such as Lexan or equivalent. Our drawing shows 3/8-inch bent baffles. If this device will be used primarily for exploding components, consider increasing the baffles to 1 inch.

2. Fabricate the HV shield as shown in Fig. 4-3, again from 1/16-inch clear polycarbonate. Note the cut slot for insertion of the insulating TRIGARM because this must allow proper alignment between the spring-loaded contacts during insertion and after removal.

3. Fabricate the positive and negative contact straps NEGSTRAP and POSSTRAP as shown in Fig. 4-4. Note that the POSSTRAP must provide proper alignment of BALL5 with BALL75 in the final assembly. Also, do not use

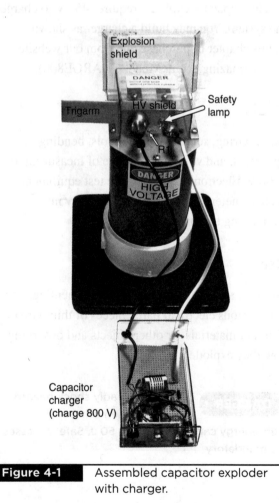

Figure 4-1   Assembled capacitor exploder with charger.

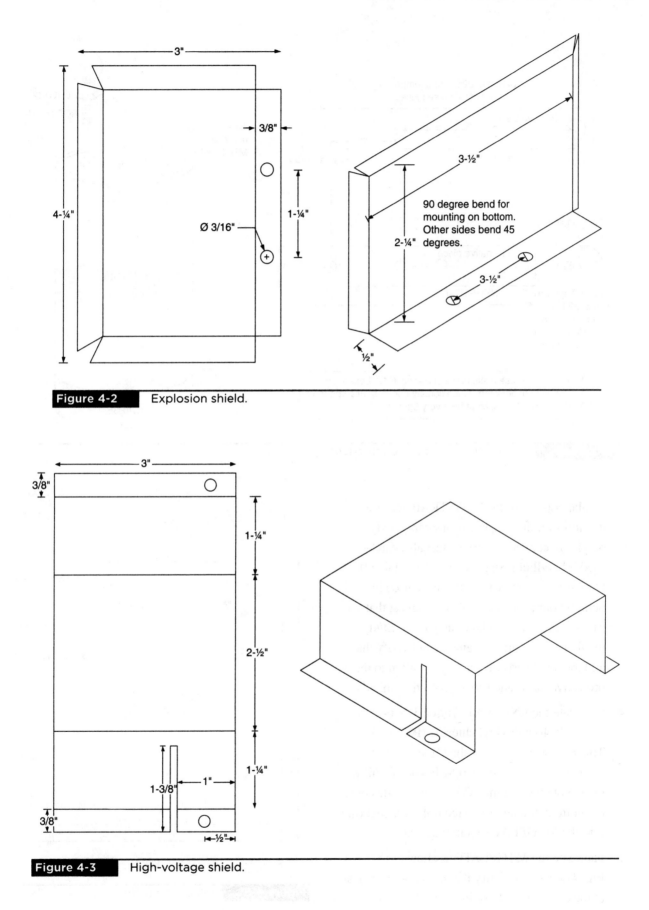

Figure 4-2   Explosion shield.

Figure 4-3   High-voltage shield.

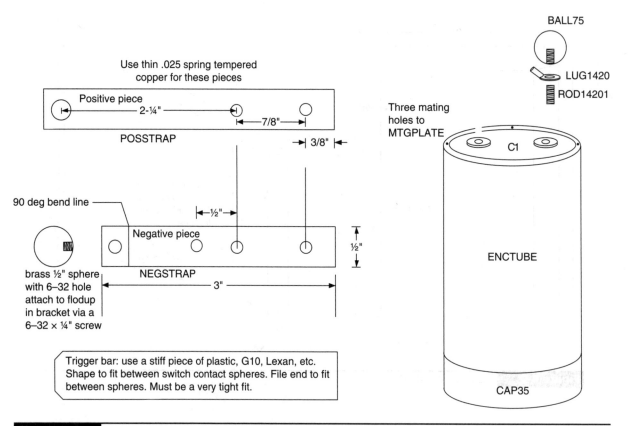

Figure 4-4 Straps, trigger arm, and enclosure.

regular copper for the NEGSTRAP because it is too soft; instead, use spring-tempered beryllium copper (or spring steel also can be used if beryllium copper cannot be obtained). Dimensions as shown are those we used in building our prototype unit. We suggest that you trial fit parts and drill holes as you go. Study the figures for parts placement, and verify that all fabricated parts match. Pay attention to the protective blast- and shock-protection shields.

4. Fabricate the ENCTUBE from a 3- × 6-inch Schedule 40 polyvinyl chloride (PVC) tube. This nicely sleeves over the C1 storage capacitor (such as the 6300-µF, 400-V unit we use). Note the 3½-inch CAP35 that slides over the bottom end and the three holes for securing it to the MTGPLATE mounting plate.

5. Fabricate the MTGPLATE as shown in Fig. 4-5. Note to verify the three mating holes at the edges with those in the ENCTUBE.

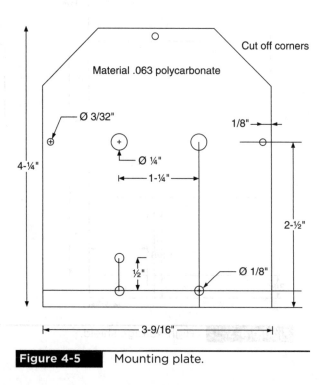

Figure 4-5 Mounting plate.

6. Fabricate the TRIGARM as shown in Fig. 4-4. This piece prevents the contact balls from touching during the charging cycle. It is quickly removed for discharging into the target load.

7. Proceed to assemble the unit by securing the 3.5-inch CAP35 to the BASE with two wood screws. Insert the ENCTUBE into the CAP35.

8. Fasten the POSSTRAP and NEGSTRAP from Fig. 4-4 to the MTGPLATE from Fig. 4-5 as shown in Fig. 4-6. Use the three SW63225s and NUT6s. Note that the two outer screws also secure the EXPSHEILD from Fig. 4-2.

9. Position the assembly from step 8 onto the capacitor using the ¼-20 brass screw SW63225 for the POSSTRAP connecting to the positive (+) terminal of C1. Include LUG1420W under the screw as shown.

10. Connect the negative (−) terminal of C1 to contact ball BALL75 using a piece of ¼-20 brass rod. Include LUG1420W under the screw as shown. Wire in the safety circuit consisting of NEON and R1 connected in series across the C1 terminals. This is shown in detail in Fig. 4-6. It should be mentioned again that for safety purposes, when the neon safety lamp is glowing, there is still 70 to 90 V on the capacitor, and caution must be used.

11. Fasten contact ball BALL5 to the bent flange of the NEGSTRAP using a SW63225 screw. Shape it so that there is a tight yet spring-like contact between the two contact balls. These are separated by the insulating TRIGARM and act like a switch when the TRIGARM is removed, allowing the ball to make a very tight and positive contact.

12. Slide the assembled C1 into the assembly from step 7. Note the three 3/32-inch holes along the edge of the MTGPLATE (see Fig. 4-5). These holes are for securing the assembly to the ENCTUBE by means of the three SW625 sheet-metal screws. Two of the holes secure the HVSHIELD from Fig. 4-3. Once this is in position, you can drill the pilot holes for the screws.

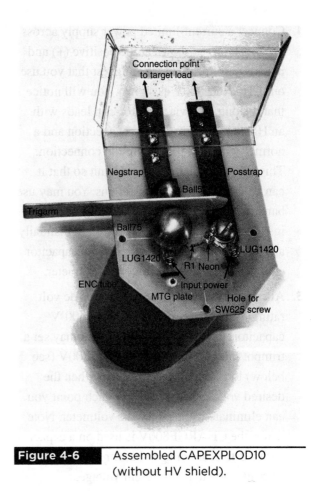

Figure 4-6　　Assembled CAPEXPLOD10 (without HV shield).

Operation

1. Verify that the switch that controls the contact balls has a tight spring-loaded pressure and that the TRIGARM can be inserted between the contacts and removed quickly, allowing the contact balls to snap together, providing a positive and secure contact. Contact bounce is minimized by the difference in weight of the contact spheres. Be sure to insert the TRIGARM before charging the capacitor.

2. **Verify that the NEON in the safety circuit (step 10) is not glowing, which would indicate that a charge is still on C1.**

3. Connect the target load to the end holes in the NEGSTRAP and the POSSTRAP. You can secure these connections with screws and nuts or even clothes pins. It is important that a good positive contact is made.

4. Connect a current-limited power supply across the capacitor, as shown at the positive (+) and negative (−) locations. We suggest that you use our CHARGE800V as shown. You will notice that the output is via alligator-clip leads with an HV lead for the positive connection and a normal-wire lead for the ground connection. This provides flexibility to the unit so that it can be used for other applications. You may use banana jacks and plugs for a more permanent installation. This system is controlled manually, and we suggest that you monitor the capacitor charge voltage using an external voltmeter.

5. Allow charging up to the capacitor's dc volt rating (in our case, with a 6300-uF/400-V capacitor, this would be 400 V). You may set a trimpot internally in the CHARGE800V (see below) to illuminate a small light when the desired voltage is reached, at which point you can eliminate the need for the voltmeter. Note that if the CHARGE800V is used on a capacitor with a different voltage, the trimpot will need to be readjusted to the different voltage.

6. **Verify that all onlookers have on safety glasses, and pull out the TRIGARM, noting an explosion as the part under test disintegrates. A thin aluminum wire actually will detonate!**

CHARGE800V

A specialized capacitor charger is needed to power up and put energy into your capacitor. This handy electronic circuit is the device that we use, a manually programmable charger for electrolytic, photo flash, and equivalent capacitors from 100 to 800 V at up to 30 mA, and it does the job nicely (Fig. 4-7). Recommended capacities are between 100 and 5000 μF, which equates to over 3000 J! (Note that the kinetic energy of a high-powered 0.30-06 rifle round is 750 J.) Units are manually voltage-controlled by an internal adjustment that illuminates a charge-indicator lamp when the required voltage is reached. This feature helps to

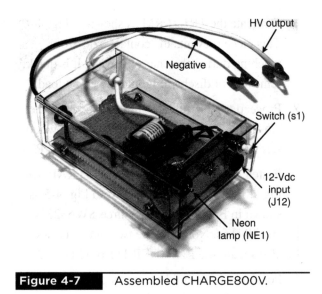

HV output

Negative

Switch (s1)

12-Vdc input (J12)

Neon lamp (NE1)

| Figure 4-7 | Assembled CHARGE800V. |

prevent overcharging and potentially dangerous explosions. Charging is current-controlled by our unique circuitry. Input power is 12 Vdc, which can come from C- or D-cell batteries, a dedicated 12-V battery, or a 12-Vdc adapter for use with 115-Vac wall power. Charging rate is over 25 (W-s)J. Size is relatively compact at $5 \times 4 \times 2$ inches, and the device can be made smaller if desired.

CHARGE800V Operating Instructions

DANGER Do not use this unit unless you fully understand the associated hazards of working with high voltage.

DANGER A serious deadly shock hazard exists when working with high-energy capacitors above 50 J.

Joulian Formula

Calculate joules by squaring the charge voltage, multiplying by half the capacitance in microfarads, and dividing by 1 million (to compensate for the measurement in microfarads). If the result is over 50 J, use extreme caution because improper contact can electrocute or cause serious burns.

$$\text{Joules} = \frac{(\text{charge voltage})^2}{2,000,000}$$

CHARGE800V Operating Steps

1. Select a capacitor, and use the preceding formula to calculate the joules for determining if the device is hazardous. Always verify that the capacitor is discharged. You can use an insulated screwdriver for small electrolytic capacitors or a discharge-resistor shorting wand. Larger capacitors usually will have a shorting wire across the terminals.

2. Connect the leads to the capacitor, and match polarity, if any (electrolytic capacitors are polarized).

3. Connect a proper-range voltmeter across the capacitor to monitor charging voltage.

4. Connect a 12-V, 0.3-amp wall adapter or 12 Vdc battery pack to the dc jack.

5. Push and hold the charge button, and note the voltage building up on the capacitor. Do not allow the capacitor to charge beyond its volt rating, as indicated on the voltmeter. Obviously, larger-value capacitors will require a longer charging time.

6. You may now set the neon lamp trip point by adjusting the orange trimpot inside the unit. This adjustment provides an indication of charge voltage and is set to meter. This feature is handy only if you are charging to a certain voltage multiple times, such as charging several electrolytic capacitors to 450 V. If you want to now charge a photo-flash capacitor to 360 V, you must readjust or use the meter.

7. There is no more data on the handling and application of the charged capacitor or on what it is to be used for. You are on your own, and we assume that you understand the hazards, as stated earlier.

DANGER A serious deadly shock hazard will exist when you use this device with high-energy capacitors above 50 J.

CHARGE800V Assembly

Fabricate a chassis box by cutting and bending a plastic sheet (Fig. 4-8). The holes don't need to be placed exactly but only relatively to position the switch, neon lamp, and 12-V input jack; the center hole is for a nylon screw that goes through the FET heat sink and helps to secure the perf board and its circuitry.

Then solder the electrical components onto a standard perf board about 2½ × 3 inches in size (see Fig. 4-11). Figure 4-9 shows the electrical schematic of the circuitry wiring. Remember to adjust R11 so that it lights up NE1 when voltage reaches 400 V.

Figure 4-10 shows the wave shapes of the charging cycle.

Figure 4-11 shows the components and wires soldered to a standard perf board. Cut a piece of aluminum (about 1¼ inches high × 1½ inches wide) to be used as a FET heat sink, drill a central hole, and hold the aluminum in place with a nylon bolt and a standard metal locknut (Figs. 4-11 and 4-12). Also drill a hole in the perf board that will extend through the plastic chassis for another nylon bolt to secure the perf board in place.

CHARGE800V Board Wiring Connections

WR1: red wire going to front-panel switch S1 (see Fig. 4-13)

WB12: black wire going to outer sleeve of front-panel jack J12 (see Fig. 4-13)

WN1: green wire going to front-panel neon light NE1 (see Fig. 4-13)

WN2: green wire going to front-panel neon light NE1 (see Fig. 4-13)

HV-W: white HV output wire

HV-B: black negative wire

Figure 4-12 shows how these components are wired together. Note that the locknut holding the aluminum heat sink to the FET stays in place and also acts as a spacer when this board is mounted to the chassis (providing a little room between the aluminum heat sink and the front panel through which air can move), so another locknut is used to secure the nylon bolt to the front panel (see Fig. 4-7).

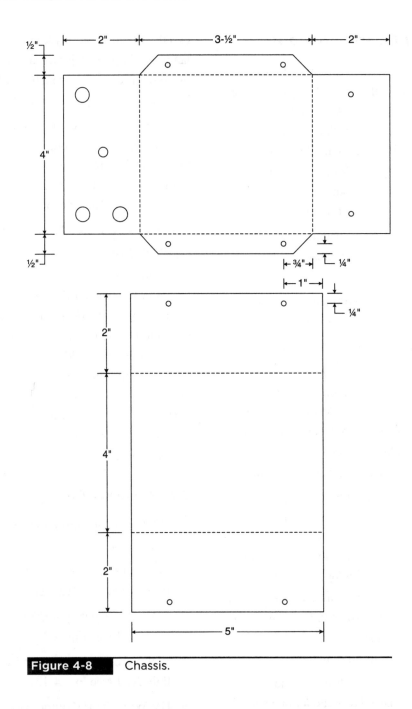

Figure 4-8   Chassis.

And Fig. 4-13 shows the front-panel components and wiring. Notice the central hole for the nylon bolt coming from the FET/heat sink.

CHARGE800V Front-Panel Connections

R12: wire going between switch S1 and the inner post of jack J12

With the perf board secured to the chassis, put the top cover on, and your CHARGE800V is ready for use.

CHARGE800V Test Procedure

First, make a simple test jig of two capacitors (here we use a 360-V, 800-µF photo flash) in series, and then "bleed" resistors (2 kΩ, 10 W) in parallel with a discharge lever to complete the circuit (Figs. 4-14 and 4-15). The components are glued to a plastic base (there is no circuitry or wiring running under the base), the discharge lever is held to the base with two screws and nuts (as shown), and the red wire cap is secured

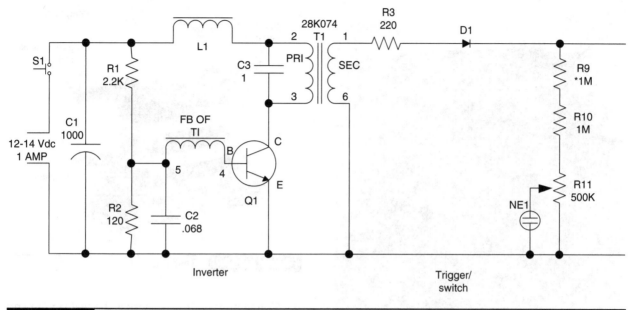

Figure 4-9 CHARGE800V schematic.

to the copper lever arm with a screw from beneath.

NOTE Our test jig uses 2 × 360-V capacitors for a total working voltage of 720 V. The CHARGE800 is capable of putting out in excess of 800 V, so be sure to monitor the voltage with a voltmeter and keep it

from exceeding your test jig's maximum, or the capacitors may blow.

Connect the safety lead to the positive clip before and after using this test jig, as well as during

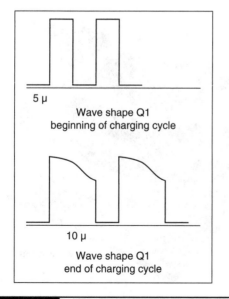

Figure 4-10 CHARGE800V charging-cycle wave shapes.

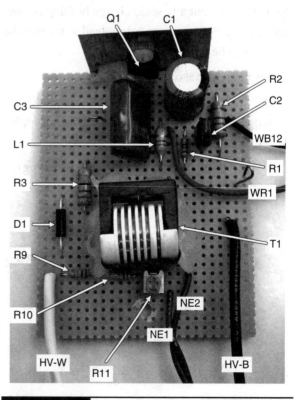

Figure 4-11 CHARGE800V perf-board component layout.

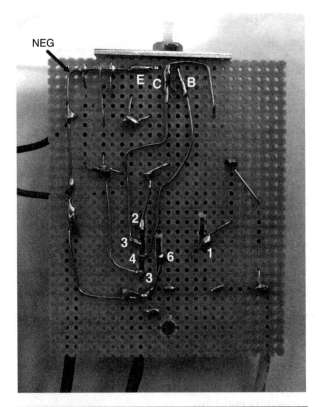

Figure 4-12   CHARGE800V perf-board component wiring.

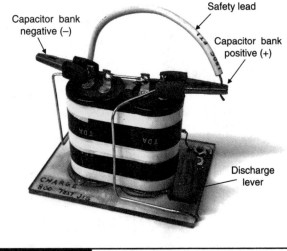

Figure 4-14   Test jig, front.

4. Connect the voltmeter across the test rig (+) and (−) (1000-V range on meter) (Fig. 4-16).

5. Plug in the wall adapter to the ballast box (green position) (see Chap. 21).

6. Press the CHARGE button, and the scope should read as in Fig. 4-17.

7. Watch the voltmeter/charge, and adjust the trimpot to the desired voltage limit. It is best to stay below 90 percent of the capacitor's maximum voltage rating. *Do not exceed the voltage rating of the capacitor load under test or it may explode and cause injury!*

storage, to prevent any static charge buildup on the capacitors. Disconnect the safety lead whenever the test jig is to be used.

1. Set your scope to 5 μs at 50 V.

2. Put the test clip on the HEAT SINK of 30SS for scope probe, and ground it to the power supply plug ground (black wire) (Fig. 4-16).

3. Connect the output to the capacitor bank (test jig): white to positive (+), black to negative (−) (Fig. 4-16).

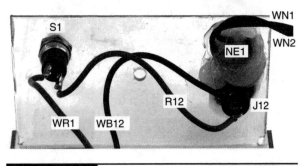

Figure 4-13   Front panel (inside view).

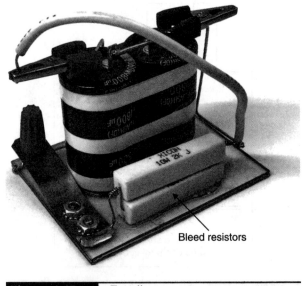

Figure 4-15   Test jig, rear.

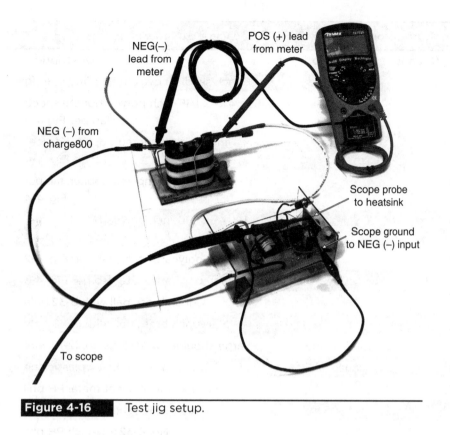

Figure 4-16   Test jig setup.

8. Discharge the test jig by pressing down the RED CAP on the discharge lever (Fig. 4-18). Watch the voltmeter until it goes to zero volts, and then take the white grounding wire and clip it to the positive (+) side of the cap. Stick it through the hole in the alligator clip.

Monitor the capacitor voltage by keeping the meter attached. Be sure to press the discharge lever and hold it for a few seconds *before and after* using the test jig so that the capacitive load is fully discharged.

This is how we test the CHARGE800; you can use your own capacitive load if you wish. As always, caution is advised because this 800-V capacitor can cause a good jolt.

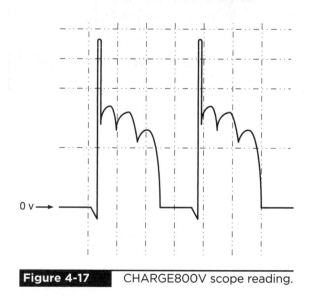

Figure 4-17   CHARGE800V scope reading.

Figure 4-18   Discharging the test jig.

TABLE 4-1 Parts List for the CAPEXPLOD10 Capacitor Exploder

Info Part	Quantity	Vend. No.	Description	Other	Price
Enctube			3- × 6-inch sked 40 pvc cut to 6 inches long		
Mtgplate			3-19/16- × 4¼- × 1/16-inch polycarbonate sheet; fab per Fig 4-2		
Posstrap			½- × 4-inch thin tempered-copper; fab per Fig. 4-4		
Negstrap			½- × 3-inch thin tempered-copper; fab per Fig. 4-4		
Trigarm			1- × 4- × 1/16-inch polycarbonate sheet; fab per Fig. 4-4		
Expshield			3- × 4¼-inch polycarbonate; fab per Fig. 4-2		
Hvshielfd			5-7/8- × 3-inch polycarbonate; fab per Fig. 4-3		
Ball5			½-inch brass ball with 6-32 hole		
Ball75			¾-inch brass ball with ¼-20 hole		
Lug1420w	2	Mou 517-1219	Large shoulder ¼-20 lugs no. 12-10 wire		
Cap35		Alliance a3.5	3½-inch black plastic cap		
SW625	3		No. 6 × ¼-inch sheet metal PH phil		
SW812	2		No. 8 × ½-inch sheet metal type B		
SW63225	4		No. 6-32 × ¼-inch PH phil		
NUT6K	3		No. 6 KEP nut		
SW14205			No. 1/4-20 × ½-inch brass screw, RH slot		
ROD14201			No. 1/4-20 × 1-inch threaded rod		
C1			6300 µF at 400 V electrolytic capacitor	6300 µF/400 V	
Neon			Small neon indicator lamp with leads		
R1			1-MW ½-W resistor		
Base			8- × 8-ft piece of ½-inch plywood, finish and paint for aesthetics		
12 DC/.5		Optional	12-V 0.5-amp wall adapter		
Charge800V		Optional	800-V charger	CHARGE800V	
		Optional	12-V battery pack AA with mating plug		

Ref No.	Quantity	Description
TABLE 4-2 Parts List for CHARGE800V Capacitor Charger		
R1		2.2-kΩ ¼-W film resistor (RED, RED, RED)
R2		120-Ω ¼-W film resistor (BRN, RED, BRN)
R3		220-Ω ¼-W film resistor (RED, RED, BRN)
R9, 10	2	1-MΩ ¼-W film resistor (BRN, BLK, GRN)
R11		500-kΩ trimpot
C1		1000-µF 25-V vertical electrolytic
C2		0.068-µF 100-V polypro
C3		1-µF 250-V polypro
D1		6-kV avalanche diode
L1		3.6-mH inductor
Q1		1N3055
T1		28K074
NE1		
S1		
J12		3.5-mm dc jack
		Rubber feet
		Chassis 5 × 8 × 1/16 inch
		Cover 5¼ × 7½ × 1/16 inch
		Heat sink 1¼ × 1½ × 1/16 inch
		Perf board 2½ × 3 inches

CHAPTER 5

Field Detector (Ion, Electrical Field, Lightning, and Paranormal Disturbance Detector)

Overview

Provides a low-cost method of detecting selectable electrical fields. This device detects both positive and negative electrical fields produced by ions, ultralow static fields, lightning, and other associated phenomena.

Hazards

No real hazards exist, with the exception of using this device during electrical storms or around other high-voltage sources (and this is just the physical hazard of holding a metal object under these conditions). Eye protection should be worn when making and testing this device.

Difficulty

Requires intermediate skills in wiring and soldering, along with certain basic sheet-metal work to form and fabricate sheet-metal chassis, cover, and so on. Or you may be able to purchase a chassis/electrical box with similar dimensions from electronics supply houses such as Mouser or Digi-Key, at which point all you need to do is drill the holes.

Tools

Basic wiring, soldering, hand tools, bending apparatus, and voltmeter capable of measuring to 40 kV. Electronic ability to use test equipment such as a volt/ohm/milliamp meter is necessary for completing this project.

Explanation of Controls (Fig. 5-1)

METER SENSITIVITY: turns on/off and adjusts meter sensitivity level. Try to keep this set for the highest sensitivity, or full clockwise (FCW).

RESPONSE TIME: may take getting used to because when the device is used in weak electrical fields, it may give erratic readings until things settle down. There is also a certain amount of meter drift in this high-sensitivity mode that is attributed to semiconductor diode drift with temperature and circuitry drift with humidity. We suggest that you try to keep the meter setting at midscale for good sensitivity. The circuit is built with high-resistance parts to try to minimize this effect as much as possible.

DAMP: sets the speed that the meter moves to changing electrical phenomenon. In the UP position, the meter will respond faster and may be difficult to set in high electric activity.

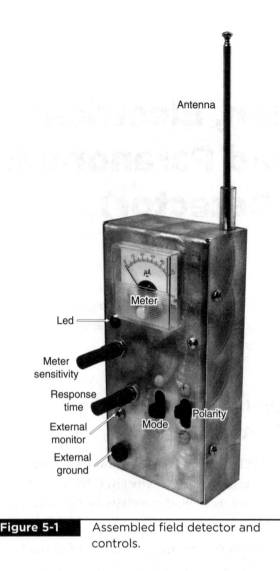

Antenna

Meter

Led

Meter
sensitivity

Response
time

External
monitor

External
ground

Mode

Polarity

Figure 5-1 Assembled field detector and controls.

EXT OUT: a 3.5-mm stereo jack that provides an ungrounded out to a 1000-Ω input for chart recorder, alarm, turn-off control, and so on.

LED: illuminates with changing brightness in strong, potentially dangerous fields.

ANTENNA: pickup probe for electrical energy; extend for weak conditions, and vice versa. In many cases you can adjust the sensitivity by adjusting the antenna height.

METER: gives a relative indication of electrical field strength in microamps.

POLARITY: switches the antenna from receiving positive (+) fields in the UP position or negative (−) fields in the DOWN position.

REMOVABLE GROUNDING LEAD: for returning the charge potential to ground (Fig. 5-2). It is not needed if you are outside or on a damp or cement floor because these places are conductive enough to virtually ground any currents accumulating on the metal enclosure by your hand contact. However, if you are inside on wooden floors you will need to clip onto preferably a known grounded point such as the *ground pin of an alternating-current (ac) receptacle.* You can use a large conductive object. The reading will be erratic and not make sense otherwise.

Ion, Charge, and Electrical Field Detectors

NOTE This device is also used by researchers investigating paranormal activity.

This very versatile tool detects both positive and negative electrical fields produced by ions and ultra-low-static fields. This unit is sensitive enough to detect the minute charge produced by just rubbing a piece of plastic and bringing the probe into proximity. It easily measures the electrical field during a thunder storm or other high-static conditions.

Figure 5-2 Ground Lead.

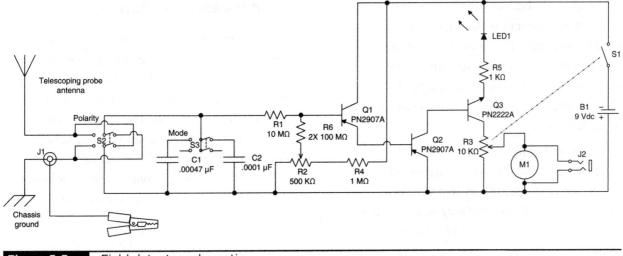

Extremely useful for mapping the electrical fields associated with high-voltage generators. Indicates dangerous E fields that may damage sensitive electronic circuits or components. Detects fields that can be dangerous to personnel. Detects the presence of weak to strong static charges from lightning, potential corona points and the output of ion generators, leaking high-voltage insulators, and wiring insulation. It is a very valuable device for the high-voltage researcher and experimenter.

The unit is built into a handheld 6- × 3- × 1½-inch metal box with extendable detection antenna probe. Simple high-performance circuitry is powered by a single 9-V battery that seldom needs replacement.

Detector indicates field strength using an analog meter with a sensitivity control. Indicates a potentially dangerous high-charge field condition with a light-emitting diode (LED). Single switch selects positive or negative charge fields. A response switch allows fast or slow response. A simple external terminal provides a connection point for a grounding lead. An RCA phono jack provides a facsimile out for remote monitoring or trigger circuitry for alarms, x, y recorders, and so on.

Remove rear cover via four screws, and insert a fresh 9-V alkaline battery. Note the Velcro pieces for securing the battery in the case. Batteries last a long time.

Note that the antenna probe on your unit is a telescoping antenna that is properly secured and electrically isolated. It is important to remember that any type of leakage around the input of Ql can reduce sensitivity. You may wish to coat the circuitry with a good-quality varnish. Make sure that the unit is dry and clean before sealing.

If you do not use a metal case, make a virtual ground with metal contact tape on the case for hand contact.

The unit is shown wired for NEGATIVE ion detection (Fig. 5-3). This may be changed to POSITIVE by replacing the PN2222 transistors with PN2907. For a quick indication, you may grab the antenna and use the body of the unit to detect the negative ions.

C1 is used to slow down the response and can be eliminated when detecting fast-discharge fields such as lightning.

Applications

Your ion/charge detector is a very, very sensitive electrical field detection device. It can be used for relative measurement but is not designed for absolute measurement. Sensitivity can be in the picoamp range when the device is stabilized.

For quick indications of the presence of a charge field, the unit is handheld and can be used to determine the location of the source. The sensitivity of this device can be realized by a simple experiment of running a plastic comb through your hair and placing the comb near the probe antenna.

If you are familiar with the metallic-leaf electroscope, you will soon realize the advantage of portability and sensitivity that this device offers. When used for indicating *or* testing the relative strength of charge/field sources, the unit should be hardwire grounded for best results. Now adjustments to a known source to determine output may be made, noting the meter reading and then readjusting the METER RESPONSE to bring meter reading on scale.

A very interesting phenomenon will be noted when you use this device to detect residual ion fields, shielding of ions, field direction, static charges, resulting polarity, and intensity of static charges, as well as a host of others. The unit is an invaluable tool determining the output of ion generators or air purifiers, the presence of dangerous static electricity, situations associated with lightning, and so on.

Many sources of charged particles soon become apparent when you use this device. People's clothing, fluorescent lighting, plastic containers, certain winds, and so on all will indicate a charge.

Please note that the antenna probe on the unit is a telescoping antenna properly secured and electrically isolated.

It is important to remember that any type of leakage around the input of Q1 can reduce the sensitivity. The input to the unit must be switched by low-leakage slider switches. The ones used we use have 5- to 10-W resistance. Unfortunately, they will reduce performance in high-humidity environments. The unit is still operational, but its performance at the superlow sensitivity may be sluggish.

Use in Paranormal Research

It has been observed that certain paranormal activity is often accompanied by a changing electrical field as the "entity" moves about. These changing fields are usually too weak to be detected by conventional detectors. The IOD40 has on several occasions detected unexplained fields coincident with other events. *At the time of this writing, a study is being done at several selected grave sites to see whether weak fields can be detected and correlated with other unexplained events.*

When used for paranormal research, detection of changing electrical fields in areas where floors are wood, tile, or covered with rugs, we suggest that you place the device on a large metal object or connect it to an earth ground such as the ground pin of an ac receptacles. If you are standing on earth ground or on other noninsulated surfaces, simply holding the unit with your hand provides the necessary virtual grounding.

Assembly

Fabricate the chassis from 1/16-inch sheet aluminum that is cut, drilled, and bent as shown in Fig. 5-4. The holes cut for the front-plate components do not need to be placed exactly but only positioned such that the components have a good layout in the box.

The back cover can be made from thinner 1/40-inch (0.025-inch) bent aluminum (Fig. 5-5).

Then secure the front-panel components and controls to the chassis (Fig. 5-6). Use nylon bolts to hold switches S2 and S3 (the four chassis holes for these switches should be tapped for 6-32, as noted in Fig. 5-4), and make a simple L-bracket from bent plastic that both holds the base of the antenna and also electrically insulates it from the chassis. Use a 2¾-inch piece of ¼-inch rubber hose to insulate the neck of the antenna from the aluminum chassis. Place a small piece (about 1 inch square)

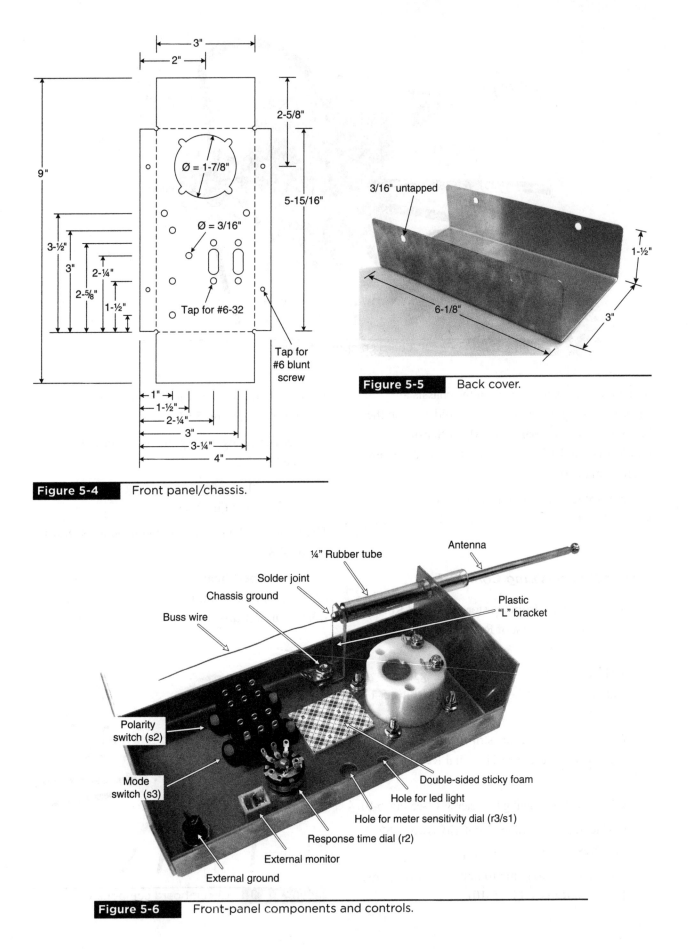

Figure 5-4   Front panel/chassis.

Figure 5-5   Back cover.

Figure 5-6   Front-panel components and controls.

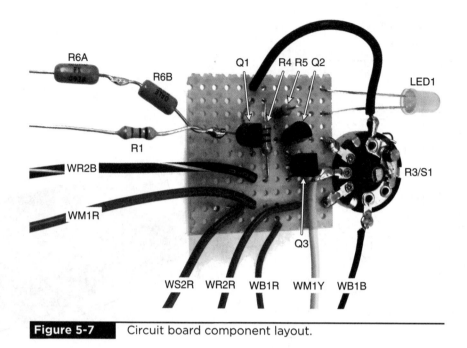

Figure 5-7 | Circuit board component layout.

of double-sided sticky foam on the chassis as shown in the figure to insulate the soldering on the bottom of the perf board. The Meter Sensitivity dial R3/S1 and LED will be installed after the perf board is made (next step).

Next, solder the electronic components onto a standard perf board that is about 1 × 1 inch in size (Fig. 5-7).

Circuit Board Wiring Connections

BR6: buss wire from resistors R6A and R6B going to second post of Response Time dial R2 (see Fig. 5-10)

BR1: buss wire from resistor R1 going to centerposts of switches S2 and S3 (see Figs. 5-11 and 5-12)

WR2B: black and white striped wire going to third post of Response Time dial R2 (see Fig. 5-10)

WM1R: red wire going to meter M1 (see Fig. 5-9)

WS2R: red wire going to Polarity Switch S2 (see Figs. 5-9 and 5-11)

WR2R: red wire going to first post of Response Time dial R2 (see Fig. 5-10)

WB1R: red wire going to positive (+) battery terminal (see Fig. 5-9)

WM1Y: yellow wire going to meter M1 (see Fig. 5-9)

WB1B: black wire going to negative (−) battery terminal (see Fig. 5-9)

The wiring of the perf board components is shown in Fig. 5-8.

With the perf board finished, remove the hold-down bolt from the Meter Sensitivity dial R3/S1, mount this to the chassis, and bolt it

Figure 5-8 | Circuit board component wiring.

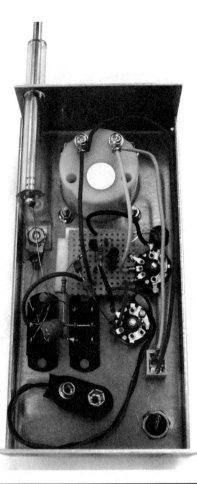

WR2R   BR6   WR2B

Figure 5-10   Response Time dial.

make contact with the antenna buss wire that crosses its path from above. Once this is done, just follow the other connections one at a time, and the device will go together easily.

When finished, apply silicone goop to the top of the antenna tube and base of the antenna,

down from the front. This will hold the perf board in place. Put the LED bracket in the chassis, and bend the LED into the bracket. Then wire all the components together, as shown in Fig. 5-9.

Detail of wiring to the Response Time dial R2 is shown in Fig. 5-10.

The wiring between switches S2 and S3 is a little less straightforward, so a detailed picture (Fig. 5-11) and schematic (Fig. 5-12) are provided to aid in assembly. The capacitor wiring is exaggerated in Fig. 5-12 to provide clarity to the wire paths and connection points. The main thing is to bend *down* the chassis ground buss wire where it travels between the top and bottom posts of polarity switch S2 so that it will not

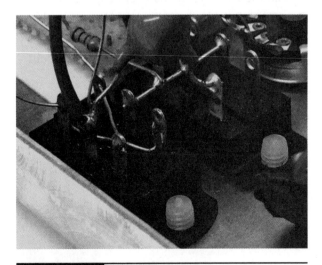

Figure 5-11   Switch wiring.

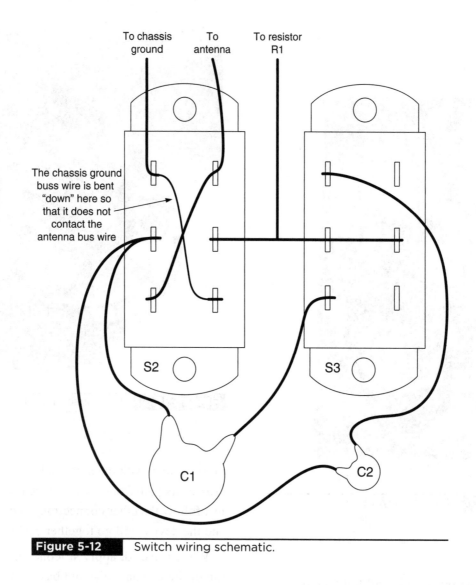

To chassis ground   To antenna   To resistor R1

The chassis ground buss wire is bent "down" here so that it does not contact the antenna bus wire

S2   S3

C1   C2

Figure 5-12 Switch wiring schematic.

around the LED light, and between some of the Damp and Polarity switch wiring to ensure that they do not make contact with each other. Let the silicone dry, install a battery (Fig. 5-13; cut a strip of Velcro tape to hold the battery, which can be found at most craft or fabric stores), and then attach the back cover. This device is now ready to use.

| **Figure 5-13** | Circuitry layout. |

TABLE 5-1 IOD40 Parts List		
Ref. No.	**Quantity**	**Description**
R1		10-MΩ ¼-W film resistor (BRN, BLK, BLU)
R2		500-kΩ 17-mm pot/switch
R3/S1		10-kΩ 17-mm pot/switch
R4		1-MΩ ¼-W film resistor (BRN, BLK, GRN)
R5		1-kΩ ¼-W film resistor (BRN, BLK, RED)
R6A, B		100-MΩ high-megaohm resistor (or 1 × 200-MΩ resistor)
C1		0.00047-µF 10-kV ceramic disk
C2		0.0001-µF 50-V ceramic disk
S2,3		DPDT low-loss slider switch
LED1		Green led
M1		50-µA 2-inch panel meter
J1		Banana jack
J2		Stereo audio jack
B1		9-V battery
		2-ft telescopic antenna

CHAPTER 6

High-Power Therapeutic Magnetic Pulser

Experiment with deep sleep, dream control, relaxation, sexual enhancement, mind control, magnetic healing, and therapy. Project uses conditioned magnetic pulses for health research, with claims of increased blood flow, cell regeneration, growth enhancement, treatment of sexual dysfunction and other related issues, and treatment of many symptoms, including pain relief.

Overview

This advanced high-powered Magnetic Pulser is purported to eliminate or reduce pain, increase cellular growth, and address many ailments. It is a very popular product that generates precision-timed magnetic pulses at selectable pulse-repetition rates and energies. Powerful enough for experimental and laboratory purposes, with pulse energies up to 1 J, but also employable for individual use. These pulses feed a coil that is applied to a target area. Pulse-repetition rates can be adjusted between 1 and 1500 Hz (from the lower alpha, beta, delta, theta, sleep, dream-control, and relaxation states to the beneficial higher frequencies)—and all at adjustable magnetic pulse energy levels, as well as choice of polarity, with a digital frequency meter to display the selected repetition rate. The unit can also generate a slowly changing pulse rate that sweeps the gamma, alpha, beta, delta, and theta ranges, inducing a sleep or highly relaxed state of mind (Fig. 6-1).

Hazards

Uses stepped-up 155- to 300-V alternating current (ac) for powering the circuit modules. Eye protection should be worn when making and testing this device.

Difficulty

Advanced. From a circuit perspective, this is one of the most complex items in this book. Requires intermediate skills in wiring and soldering, along with certain basic sheet-metal work to form and fabricate the sheet-metal chassis, cover, and so on. Or you may be able to purchase a chassis/electrical box with close enough dimensions from electronics supply houses such as Mouser or Digi-Key, at which point all you need to do is drill the holes.

Tools

Basic wiring, soldering, hand tools, bending apparatus, voltmeter, frequency meter capable of

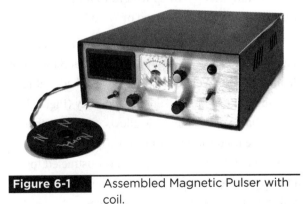

Figure 6-1   Assembled Magnetic Pulser with coil.

measuring from 10 to 2000 Hz, and a *low-cost oscilloscope*.

You should familiarize yourself with Chap. 21 before building this circuit or any similar circuit without an isolation transformer. We advise that you use the Testing Circuit Jig shown in Chap. 21 for this project because it verifies dangerous ac grounds. Other amenities include powering a virgin circuit starting with low input voltage and slowly increasing the voltage, noting that any excessive current could be dangerous and totally wipe out your hard work!

Theory of Operation

The pulser coil can be worn or held against any part of the body with your hand or kept in place with a tension bandage. A second set of connection jacks allows a second coil to be used for those who want to place the target area between two flux-producing elements or for use with another subject. You also can use this pulser while driving and on the go with a small, low-cost inverter plugged into an automobile's 12-Vdc cigarette lighter. These inverters can be bought from an electronics store, some of the larger "marts" may carry them, or you can get them on our website if they can't be found locally (www.amazing1.com).

A special circuit provides a sleep and relaxation mode that sweeps through the alpha, beta, theta, and delta frequencies (100 Hz down to 0.5 Hz) at a cycle period of 10 seconds or 10 minutes. You may set this Auto Sleep Sweep feature to repeat or just turn off after you hit the land of "nod."

The basic coil has windings that are 2 inches in diameter (the disk holding the coil is about 3 inches in diameter), and a Velcro strap can be attached to it for securing to most target points on the body (Fig. 6-1). A detachable extension cable allows other magnetic coils to be used for other functions. These coils can be made to all shapes and sizes and constructed for a variety of uses because this unit is built to operate with a wide range of output impedances.

Magnetic pulses produced electrically often contain a mixture of both north and south pole directions. This is caused by circuit ringing and decays exponentially over a period of time longer than the intended pulse-duration time. Our design uses patented circuitry that all but eliminates this undesirable effect, allowing the Magnetic Pulser to apply the north or south poles only to the body. The coil strength and polarity are measured easily with a Magnetic Strength and Polarity Meter, such as our MX3511. When using this meter, start from a distance, and do not move any closer to the coil than when you just notice meter movement, or you will blitz the meter!

An interesting example that demonstrates the magnetic pulse power is to hold an aluminum can between your ear and the pulser coil, and you will hear a distinct pulsing as the can itself becomes the opposing polarity. If you place the aluminum can on the coil and turn it up to maximum energy, the bottom of the can will become too hot to touch. This phenomenon is the result of Lenz's law. If you hold the coil in front of a cathode-ray (TV) screen, you can see lines forming on the screen. *This does not work on plasma or liquid-crystal display (LCD) TVs.* On some TVs, the lines are visible when the coil is held as far away as 3 to 12 inches. This demonstrates how the magnetic pulsing action is more powerful, and effective, than traditional magnetic therapy machines.

The pulsing coil takes magnetic therapy to a whole new dimension, with exciting breakthrough results and benefits. Permanent magnets, no matter how strong, will not produce the same results as the pulsed fields of this device, which produces induced back-electromotive force (emf) currents in the target tissue. In order to do this, you must have a high-intensity, time-varying magnetic impulse—not just a magnetic field. Magnets do have their uses, but they work in different ways with different results. Equally important is the fact that we humans have "electrochemically powered" brains: All our thoughts and perceptions consist of complex networks of electrical signals and electromagnetic fields that pulse and sweep throughout the brain. Thus it then makes sense that harmonic electrical revitalization of the brain can influence your mental state and positively alter mental effectiveness.

Currently, fixed magnets are sold for health applications and are accepted by the medical profession. These magnets produce only a very weak stationary field and cannot penetrate to the tissue depths to be as effective as pulsed devices. Many users of pulsed magnetic devices have claimed a remarkable healing effect on many ailments and afflictions. The data are both from professional to regular users.

Magnetic pulsers have been registered as a Canadian medical device, but this noninvasive pulser is not yet approved by the U.S. Food and Drug Administration (FDA).

Other Benefits

Externally applied magnetic pulses to the lymphatic system, spleen, kidney, and liver are believed to help neutralize germinating, latent, and incubating parasites of all types, helping to block reinfection. This would help to speed up the elimination of disease, restore the immune system, and support detoxification.

The movement of the lymphatic system is essential in purifying, detoxifying, and regenerating the body, supporting the immune system, and maintaining health. Normally, lymph is pumped by the movement of our body through vigorous exercise and physical activity. However, a clogged, sluggish, or weak lymphatic system prevents the body from circulating vital fluids and eliminating toxic waste, thus weakening the immune system and making us vulnerable to infections and diseases. In order to be healthy, it is essential to keep the energy balanced and fluids moving so that the body operates at its optimal healing ability. From a holistic perspective, each cell must be enlivened with its own unique

energetic frequencies and harmonic energy state and be harmonically connected to the life energy throughout the rest of the body.

The theory is that Pulsed Electromagnetic Fields can influence cell behavior by inducing electromagnetic changes around and within the cell. Improved blood supply increases the oxygen content, activating and regenerating cells. Improved calcium transport increases the absorption of calcium in bones and improves the quality of cartilage in joints, helping to decrease pain. Acute and even chronic pain may decrease or even disappear.

As we get along in years, hormone production drops off and contributes to our "feeling older." These magnetic therapy devices use complex energy pulses of magnetic waves to attempt to stimulate certain body functions and accelerate the production of important hormones to revitalize and rejuvenate our physical condition. Magnetic pulsing is believed to aid in human growth hormone and neurotransmitter production. Some of the resulting claims have been remarkable, including an increase in vitality, sexuality, male penis size, and cognitive effects such as learning and reduction of memory loss, with even reports of increased psychic ability and increased IQ for some people.

Users have claimed faster healing of injuries, including bone fractures, and reductions of carpal tunnel and drug-free arthritis pain. The pulsed magnetic field is believed to stimulate blood flow and cellular respiration in the areas where it is applied. There have been reports of reductions or cessation of migraine headaches after magnetic therapy is applied.

Addendum

This device also can be used for research into disorientating and confusing animals and insects that use the earth's magnetic field for behavioral patterns, navigation and migration, mating, and other functions. Some researchers claim to have observed effects on plant growth using certain frequencies and power levels.

Certain combinations of polarity and pulse-repetition rates also have been said to disorient and confuse people. Use caution in attempting to employ the pulser for these applications.

The following list of pulse frequencies was taken from www.elixa.com/light/nogier.html and is not a result of our own research. These frequencies are claimed to be used as healing— and experimental—tools in pulsed magnetic force therapy. *We make no claims as to the veracity of this information, and this is not a complete list because research is ongoing.*

Normalize adrenal function: 1335 Hz

Normalize pituitary function: 635 Hz

Stimulate increased/normalized human growth hormone production (pituitary): 1725, 645, 1342 Hz

Stimulate normal pineal function: 480 Hz

Stimulate normalized hypothalamus function: 1534, 1413, 1351 Hz

Normalize endocrine system function: 1537 Hz

Stimulate/normalize immune system function: 835 Hz

Stimulate normal colon function: 635 Hz

Stimulate normal thyroid function: 763 Hz

Normalize progesterone levels (in sequence): 763, 1446, 1443, 763 Hz

Normalize estrogen production levels (male and female): 1351 Hz

Normalize testosterone production (male): 1444 Hz

Normalize testosterone production (female): 1445 Hz

Stimulate normal pancreas function: 654 Hz

Stimulate normal liver function: 751 Hz

Stimulate normal kidney function: 625 Hz

Stimulate normal heart function: 696 Hz

Normalize blood pressure: 15 Hz

Stimulate normal nervous system function: 764 Hz

Stimulate normal lymph system function: 676 Hz

Stimulate increased lymph system circulation: 15 Hz (15.2 Hz)

Stimulate normalized blood circulation: 337 Hz

Stimulate increased blood flow/circulation: 17 Hz

Normalize red blood cell production: 1524 Hz

Normalize white blood cell production: 1434 Hz

Normalize hemoglobin production: 2452 Hz

Stimulate the reinforcement of DNA integrity: 528 Hz

Stimulate the reinforcement of RNA integrity: 637 Hz

Stimulate clarity of thought/mental function: 35 Hz

Stimulate the stabilization of emotional states: 15 Hz

Stimulate the clearing of emotional trauma/energy blocks: 15 Hz

Stimulate the balancing of spiritual well-being: 1565 Hz

Reduce chemical sensitivity: 443 Hz

Reduce electrical sensitivity: 657 Hz

Stimulate the normalization of calcium metabolism: 326 Hz

Stimulate repair/healing of nerve damage: 578, 764, 657 Hz (also try 2.0, 657, 5000, and 10,000 Hz)

Stimulate the healing of bones: 7.0 Hz

Stimulate the healing of ligaments: 9.7 Hz

Stimulate the healing of muscles: 13.5 Hz

Stimulate the healing of capillaries: 15.2 Hz

Reduce swelling of herniated disk: 25.4, 326, 15.2 Hz

Reduce excess fluid retention in joints and tissues: 15.2, 24.3 Hz

Reduce general back pain (fibromyalgia): 326 to 328 Hz

Stimulate endocrine production (stroke recovery, etc.): 2642 Hz

Accelerate the healing and clearing of scarring: 5.9 Hz

Accelerate the healing of injuries and surgeries (promote regeneration): 47 Hz

Reduced the effects of diabetic neuropathy: 73 Hz

Pulsed Electromagnetic Therapy is also becoming recognized by orthodox medicine and making appearances in the mainstream, such as on NBC's *The Dr. Oz Show* (outtakes of which can be seen at www.magnetictherapywerx.com, or do a YouTube search for "pulsed electromagnetic therapy").

Circuit Operation

This Therapeutic Magnetic Pulser project uses the most complex electronic circuitry in this book (Fig. 6-2). It is not recommended for the inexperienced hobbyist. Finished units ready to use are available on our website.

We build our Control Module on a dedicated printed circuit board (PCB), but it also may be built on a perf board if the wiring layout in Fig. 6-5 is followed. The control board generates and controls the variable pulses. A function generator integrated circuit (IC3) XR2206 with PULSE frequency rate determined by the remote-control (RC) timing circuit using capacitors C2 and C3 and with R21FP allows adjustment from 0.5 to 1500 Hz in two ranges as selected by the rotary multifunction switch S2BFB. A digital frequency meter (FREQMETFP) reads out the frequency selected. The high range is from 50 to 1500 Hz, whereas the low range is from 0.5 to 96 Hz.

A very important function is the FREQUENCY SWEEP mode as selected by switch S2AFP. The two positions allow a sweep function to automatically change over a range of 50 to 0.5 Hz in a 10-minute or 10-second period. *The 10-second period is usually used to verify the SWEEP function.* This function allows a person using the device to relax as the frequency drops from 50 to 0.5 Hz and then repeats if needed. The REPEAT function can be disabled by a toggle switch (S3FP) if desired and usually depends on the psyche of the user. You will note that this action exposes the user to the various mind states, such as

Gamma	>40 Hz—*highest state of alertness, bursts of insight and information processing*
Beta	14–40 Hz—*normal waking/ heightened state of awareness*
Alpha*	7.5–14 Hz—*day dreaming/deep relaxation*
Theta*	4–7.5 Hz—*deep meditation/light sleep*
Delta	0.5–4 Hz—*deep dreamless sleep*

We have reports in which various combinations of time and frequencies, with the preceding values, have influenced the types of dreams experienced by the user. Repeatability on several occasions has encouraged more controlled research. In the meantime, we receive reports that, even though undocumented, do suggest a positive relationship.

The sweeping signal is produced by the internal voltage-controlled oscillator (VCO) internal circuitry of the XR2206 function generator. In order to get a linear change in the frequency, we must supply a likewise-changing voltage to the VCO. This voltage in obtained as a linear ramp by charging a capacitor with a constant current $i(t)$. This function is provided by taking advantage of the current-producing properties of a simple transistor. Q2 is a 2907 *pnp* transistor with it base-to-emitter voltage held constant by Zener diode Z1. Base current is now I, and because the collector current is $f(\beta) \times I$, we see that the accumulating voltage on

---

*The optimal level for visualization is the alpha-theta border at 7 to 8 Hz. This is the gateway to your subconscious mind.

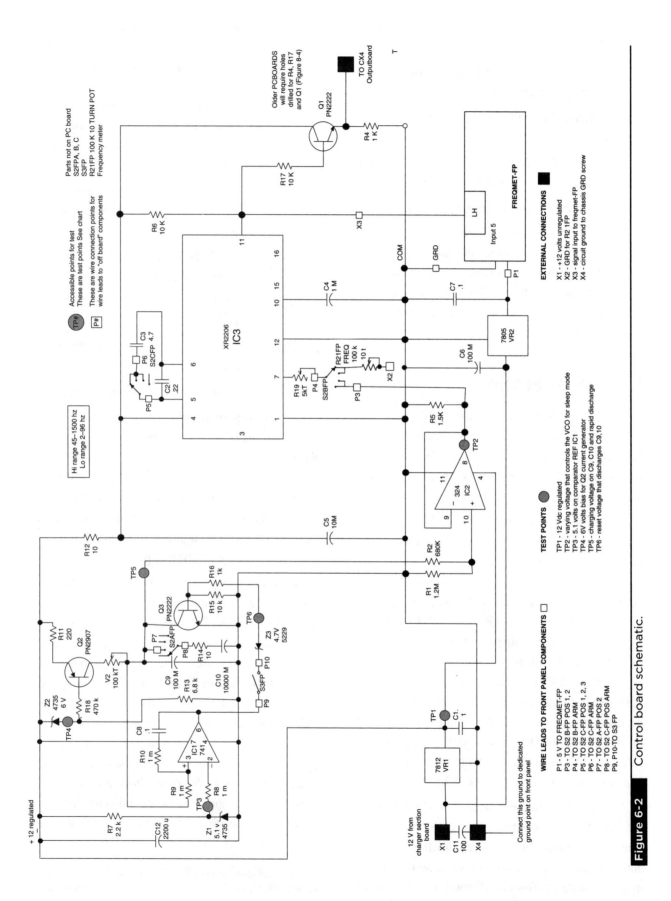

Figure 6-2 Control board schematic.

C9 and C10 is a linear ramp causing the VCO to generate a linear changing frequency as $e = it/C$. This voltage is inverted by (IC2) LM324 and fed to pin 7 of IC3. The ramp is reset by a comparator consisting of IC1 (LM741). When the ramp voltage reaches the 5.1 V set by Zener diode (Z1), it provides a voltage level that biases transistor (Q3) ON, thus discharging capacitors C9, C10 and preventing them from recharging and recycling the event. Switch S3FP must be ON for this to happen, or the event will keep repeating.

Power for the circuit is obtained through transformer T2CH, a small 12-V, 1-amp chassis-mount part. Output is rectified by diodes DX12–13. Capacitor CX2 increases the 12 V by 1.4 to 17 V, where it is connected to regulator VR1, providing a good, stable voltage. Regulator VR2 provides 5 V to the frequency meter. The meter obtains a sample of the frequency drive from pin 11 of IC3. Output transistor Q1 provides a positive-going pulse for triggering the gate of silicon-controlled rectifier SCRX1CH on the Output board (Fig. 6-3).

The Charger module section is where the 120-Vac line voltage is converted into a variable dc voltage to charge the pulse-energy capacitor C7CH. This is accomplished by a half-bridge inverter in which the rectified line voltage is converted into a high-frequency wave. This action allows further voltage step-up for supplying pulse capacitor C7CH.

The direct ac line voltage is rectified and voltage-doubled by diodes D1 and D3 and accumulated by filter capacitors C10 and C11. The 120 Vac is now converted to $120 \times 2 \times 1.4 = {\sim}340$ Vdc. The 1.4 is the peak value at the input and is capacitive to the filter/doubler network. While this system lacks regulation, it does allow more power to be delivered to the system. In-rush current is controlled by the in-rush-limiting resistor (RX) that provides a very nonlinear $V/I/t$ function. When the device is cold, the resistance is very high, limiting the peak charging current of C10 and C11 when first turned on. Once the caps are charged, the resistance drops to a value conducive of the anticipated load current.

This eliminates the large spike of current through D1 and D3 that would be a potentially damaging step transient to other components.

The rectified 340 Vdc is now fed to the half-bridge consisting of MOSFETs Q1 and Q2, where the voltage is chopped at a high frequency suitable for the primary windings of the small ferrite step-up transformer (T1CH). The input signal to Q1 and Q2 is generated by the high-side driver (IC1). It is important to note that the gate of Q1 is referenced at one-half the 340 Vdc. This is overcome by the ability of IC1 to reference its ground by boot-strap capacitor C4 fooling Q1 into a mode where its gate only sees the correct drive voltage of 15 V by the holding ability of C4 maintaining a virtual ground. Q2 is referenced from real ground and operates normally. The power to IC1 is via dropping resistors (R2, R3) coming from the midsection of the voltage-doubling capacitors C10 and C11. Even though the driver chip has a built-in Zener diode, we choose a 1-W external one as a safety. Capacitors C2 and C9 must be as close to the IR2153 as possible to supply the peak drive currents necessary for the gates of Q1 and Q2.

The Output module section is where the high-frequency, high-voltage output from T1CH is rectified by diodes DX1 through DX4 in a full-bridge configuration. This pulsating dc voltage is now integrated onto capacitor C7CH, producing the necessary charge energy where $W = CV^2/2$ J. The charge on C7CH is transferred to the capacitor discharge pulse transformer (T2xCH) via the switching action of SCRX1CH. The pulse output now drives the output therapy coil with a high peak current generating the magnetic pulse that we want. Reverse voltage across the SCR is forwarded to C7CH by reverse diode DX5. The output winding of T2xCH is connected to the four output jacks (J1–4RP) that provide connection to one or two therapy coils. They are color-coded red and black for obvious identification of coil connections. Magnetic pulses produced electrically often contain a mixture of both north and south pole directions. This is caused

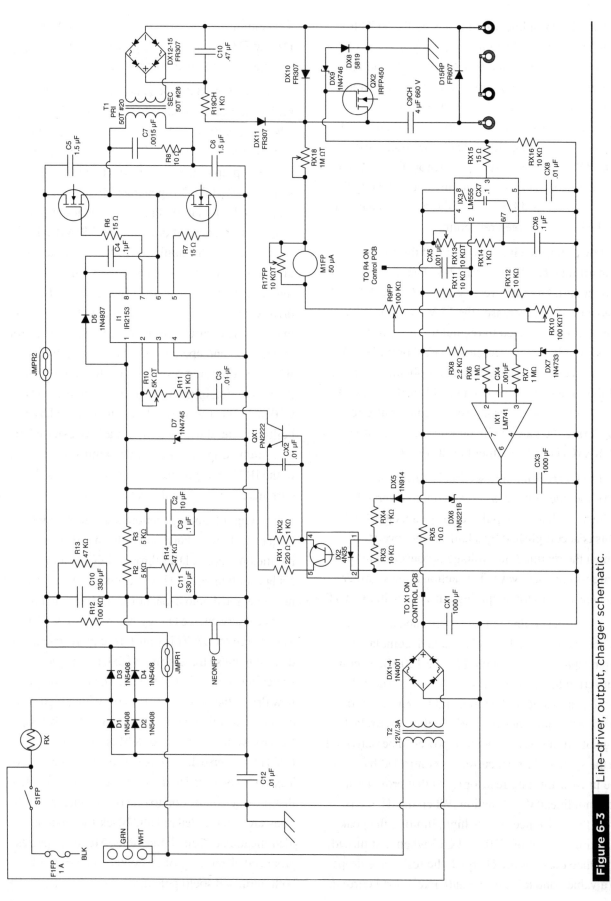

Figure 6-3 Line-driver, output, charger schematic.

80

by circuit ringing and decays exponentially over a period of time longer than the intended pulse-duration time. Our design uses a diode (D7RP) to clip the unwanted magnetic pole circuitry and all but eliminates this undesirable effect.

Assembly

Assembly of the device is divided into four parts:

1. Control module circuitry and rotary switch
2. Line-driver circuitry
3. Output/charger board circuitry
4. Chassis/component fabrication and interconnected wiring

SPECIAL NOTE Solder a standard integrated circuit (IC) socket to the board wherever ICs are installed. This allows easy replacement of ICs if (when) they burn out, which, along with MOSFETs, are the most common to fail if the circuitry is overloaded or overheated.

1. Assembly: Control Module

A PCB is used for the Control Module (Fig. 6-4), which is available on our website (www.amazing1 .com), or a perf board can be used with wiring that follows the connections on the back of the Control Module board (Fig. 6-5). Note that Fig. 6-5 is an "x-ray" view of Fig. 6-4 looking straight through to the wiring as a help to better see the connections between components. If you purchase a board, please note the additional 7 holes drilled in the right half of the board to accommodate the R3, R4, and R12 resistors and the Q4 transistor, as well as the corresponding wiring between these components.

Solder the components and wires to the board as shown in Fig. 6-6 (note that some component locations are unfilled because this board is also used for other designs). If you are using a perf board, take a marker and trace out component locations on one side and wiring connections on the other before starting the soldering.

The Control Board is wired to front-panel components and the Output/Charger Board.

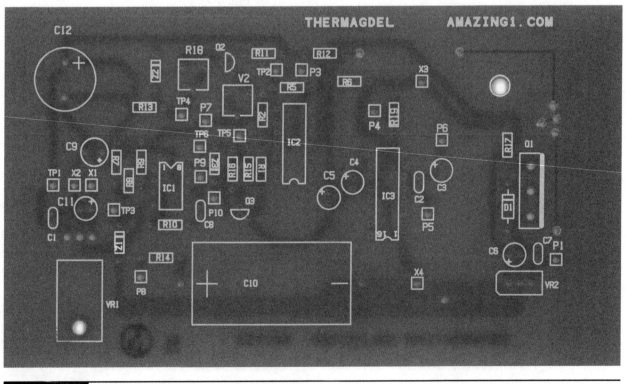

Figure 6-4   Control PCB component placement.

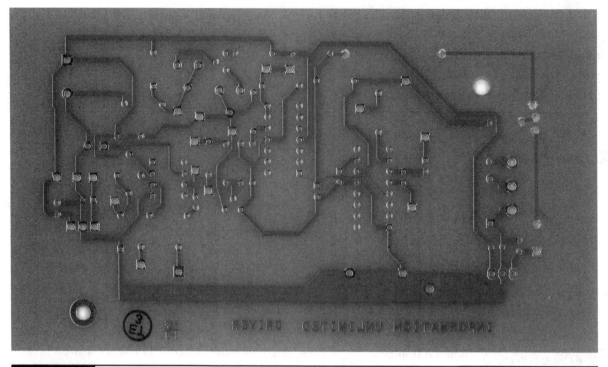

Figure 6-5 Control PCB circuitry ("x-ray" view).

Control Board Wiring Connections

P1: 7-inch red wire going to front-panel frequency meter (see Fig. 6-21)

P3: 7-inch black wire going to S2 rotary switch (see Fig. 6-7)

P4: 7-inch white and black striped wire going to S2 rotary switch (see Fig. 6-7)

P5: 7-inch green wire going to S2 rotary switch (see Fig. 6-7)

P6: 7-inch black wire going to S2 rotary switch (see Fig. 6-7)

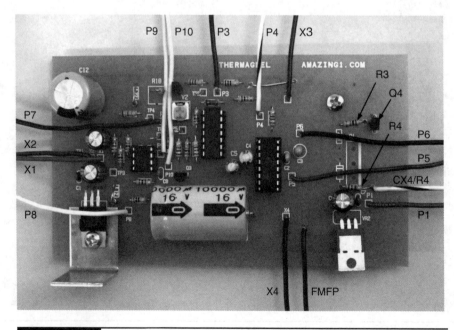

Figure 6-6 Control board, assembled.

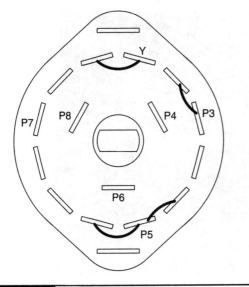

Figure 6-7 S2 rotary switch wiring.

P7: 6-inch blue wire going to S2 rotary switch (see Fig. 6-7)

P8: 6-inch white wire going to S2 rotary switch (see Fig. 6-7)

P9: 6-inch yellow wire going to front-panel S3 switch (see Fig. 6-21)

P10: 6-inch yellow wire going to front-panel S3 switch (see Fig. 6-21)

X1: 5-inch red wire going to output/charger board (see Fig. 6-14)

X2: 6-inch green wire going to front-panel R3 dial (see Fig. 6-21)

X3: 7-inch blue wire going to front-panel frequency meter (see Fig. 6-21)

X4: 9-inch black wire going to chassis ground (see Fig. 6-29)

FMFP: 5-inch black wire going to front-panel frequency meter (see Fig. 6-21)

CX4/R4: 7-inch white and black striped wire going to output/charger board (see Fig. 6-14)

Figure 6-7 shows where wires attach to the front-panel dial S2 coming from the Control Module (see Figs. 6-4, 6-6, and 6-7) plus the one yellow wire Y going to the front-panel dial R3.

Rotary Switch Wiring Connections

Y: yellow wire going to front-panel dial R3 (see Fig. 6-21) (note that the two switch terminals are soldered together)

P3: black wire going to control-board location P3 (see Fig. 6-6) (note that the two switch terminals are soldered together)

P4: white and black striped wire going to control-board location P4 (see Fig. 6-6)

P5: green wire going to control-board location P5 (see Fig. 6-6) (note that the three switch terminals are soldered together)

P6: black wire going to control-board location P6 (see Fig. 6-6)

P7: blue wire going to control-board location P7 (see Fig. 6-6)

P8: white wire going to control-board location P8 (see Fig. 6-6)

Figure 6-8 shows the rotary switch wiring.

2. Assembly: Line Driver

A PCB is used for the Line Driver (Fig. 6-9), which is available on our website (www.amazing1.com),

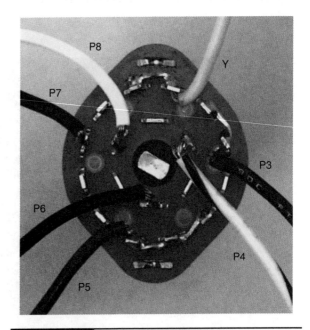

Figure 6-8 S2 rotary switch wiring.

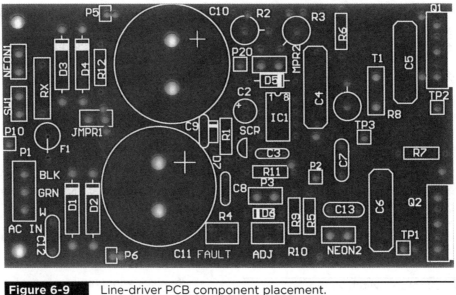

Figure 6-9   Line-driver PCB component placement.

or a standard perf board can be used with wiring that follows the connections of the PCB (Fig. 6-10). Note that Fig. 6-10 is an "x-ray" view of Fig. 6-9 looking straight through to the wiring as a help to better see the connections between components.

Solder the components and wires to the board as shown in Fig. 6-11 (note that some component locations are unfilled because this board is also used for other designs). If you are using a perf board, take a marker and trace out component locations on one side and wiring connections on the other before starting the soldering. Note the inset in Fig. 6-11, showing the jumper connection that needs to be made above the R2 and R3 resistors.

The Line Driver is wired to front-panel components, two chassis transformers, and the Output/Charger Board.

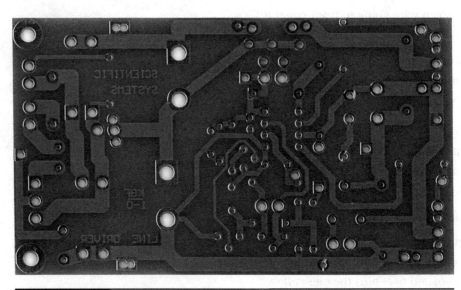

Figure 6-10   Line-driver PCB circuitry ("x-ray" view).

Figure 6-11   Line-driver PCB, assembled.

Line-Driver PCB Wire Connections

NEON-R: red wire going to front-panel NEON light (see Fig. 6-21)

NEON-B: black wire going to front-panel NEON light (see Fig. 6-21)

S1-FP: black wire going to front-panel switch S1 (see Fig. 6-21)

GND-FP: green wire going to chassis ground (see Fig. 6-29)

PWR-W: white wire going to splice with rear-panel PWR cord (see Fig. 6-29)

IX2/R2: red wire coming from resistor R2 and going to output/charger board (see Fig. 6-14)

QX1/R11: yellow wire coming from resistor R11 and going to output/charger board (see Fig. 6-14)

GND-LN: black wire going to output/charger board (see Fig. 6-14)

T1: insulated magnetic wire coming from 50-turn side of chassis transformer T1CH (see Fig. 6-29)

SPECIAL NOTE For TO247 MOSFETs, we have found that the Mouser (www.mouser.com) terminal blocks (Part No. 158P02ELK508V3-E) work perfectly as a MOSFET socket (Fig. 6-12). You don't have to use these, but they will greatly simplify things if (when) a MOSFET burns out.

The MOSFET terminals are then bent 90 degrees and secured to the aluminum heat sink (see Fig. 6-26) with screws and thermal tape to

Figure 6-12   Mouser *p/n* 158P02ELK508V3-E makes a good MOSFET socket.

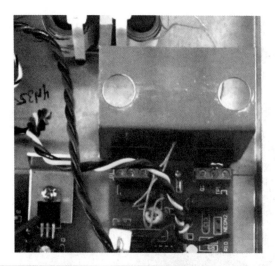

Figure 6-13 MOSFETs in their brackets.

increase heat dissipation (Fig. 6-13). They can be replaced easily if they do happen to burn out.

3. Assembly: Output/Charger Board

The Output/Charger schematic is shown on Fig. 6-3. If you are using a perf board, assemble the components in Table 6-3 as shown in Fig. 6-14, wired as shown in Fig. 6-15.

The Output/Charger Board is wired to everything: front-panel components, Control Board, Line-Driver PCB, and all chassis components.

Output/Charger Board Wire Connections

R7R: red wire going to front-panel R2 Gauss switch (see Fig. 6-21)

R7B: black wire going to front-panel R2 Gauss switch (see Fig. 6-21)

T2: red wire going to chassis transformer T2CH (see Fig. 6-29)

T2: red wire going to chassis transformer T2CH (see Fig. 6-29)

T2x: heavy (No. 14) magnetic wire coming from chassis transformer T2xCH (see Fig. 6-29)

T1: insulated magnetic wires coming from 300-turn side of chassis transformer T1CH (see Fig. 6-29)

C1x: 2-inch No. 12 buss wire going to chassis ground

IX2/R2: red wire going to line-driver PCB

QX1/R11: yellow wire coming from transistor QX1 and going to line-driver PCB (see Fig. 6-11)

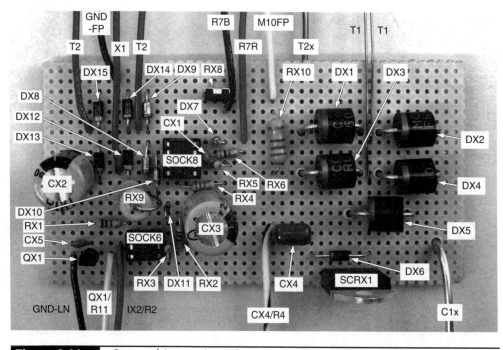

Figure 6-14 Output/charger board, assembled.

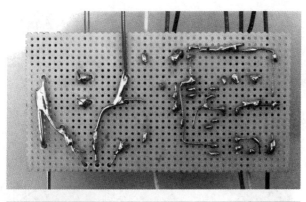

Figure 6-15 Output/charger board circuitry.

GND-LN: black wire going to line-driver PCB ground (see Fig. 6-11)

X1: 5-inch red wire going to control board (see Fig. 6-6)

CX4/R4: white and black striped wire going to control board (see Fig. 6-6)

GND-FP: black wire going to chassis ground (see Fig. 6-29)

M10FP: yellow wire going to front-panel Gauss meter (see Fig. 6-21)

Figure 6-15 shows the Output/Charger Board circuitry. Note that Fig. 6-15 is not an "x-ray" view, but it shows the actual wiring on the back of this board because this is how you will be wiring the back of the perf board.

4. Assembly: Chassis and Components

Fabrication drawings of the main chassis, cover, heat sink for Q1 and Q2, and heat sink for VR1 and VR2. Front and rear panel and chassis with their own bill of materials with part numbers having a prefix "FP,RP."

4a. Chassis and Components Assembly: Transformer T1CH

Transformer T1CH will need to be fabricated. Start by cutting two plastic tubes of the correct length and diameter (1-5/8 inch long × 5/8 inch outside diameter); then make the appropriate number of turns (Figs. 6-16 through 6-21). On one tube, use 26-gauge magnetic wire with 300 windings. On the other tube, use 20-gauge Litz wire with 50 windings. (Litz wire will give maximum efficiency, but heavy Themaleze magnetic wire will work.) On the ferrite core, place pieces of Mylar tape to act as air gaps between the two halves (Fig. 6-16).

Lay the 26-gauge magnetic wire across a tube with several inches overhanging the edge (enough to reach the component to which the coil will be soldered plus a little slack), and wrap the end of the tube with some tape to secure the wire in

Figure 6-16 Transformer components.

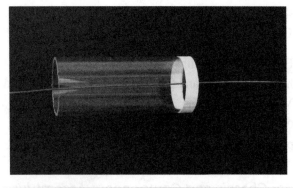

Figure 6-17 Start of transformer coil winding.

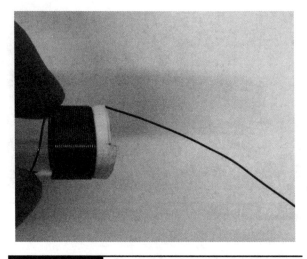

Figure 6-19 Keep the wire snug.

place (Fig. 6-17; note that the wire shown in this figure is 18 gauge, and the 26-gauge wire to be used later will be much thinner than this). If you don't have thin tape, just cut a strip about 3/8 inch wide. It is important to use a thin strip of tape here because the bend of the wire should be close to the tube edge.

Bend the wire, and start wrapping it around the tube (Fig. 6-18).

Keep the wire snug as it is wound (Fig. 6-19). Also keep track of the number of turns, either counting as the wire is wound or waiting until the end is reached and then counting—whichever works best for you.

Once the wire gets to about 3/8 inch from the other end of the tube, wrap the wire with Mylar tape to hold it in place. Be sure that the exact number of windings has been counted before it is wrapped because the wires are more difficult

to count once the coil has been taped over. Then continue winding back down the tube in the other direction until the total number of windings has been reached, and tape things securely together so that the wire does not unwind. Figure 6-20 shows a thin piece of Mylar tape securing the end of the first layer of winding (step 3), where the magnetic wire can be seen changing direction (step 4) and continuing back down the tube.

Winding the 50 turns of heavier Litz wire is the same procedure but with one difference: After the first layer of wire is wound up the coil, the winding should be completely wrapped with

Figure 6-18 Bend wire and begin winding.

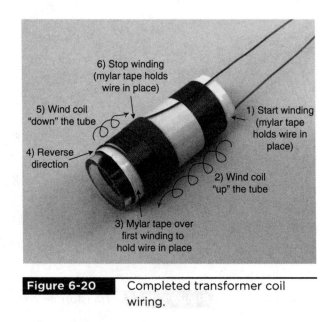

6) Stop winding (mylar tape holds wire in place)

5) Wind coil "down" the tube

4) Reverse direction

1) Start winding (mylar tape holds wire in place)

2) Wind coil "up" the tube

3) Mylar tape over first winding to hold wire in place

Figure 6-20 Completed transformer coil wiring.

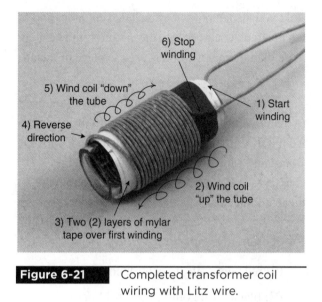

Figure 6-21 Completed transformer coil wiring with Litz wire.

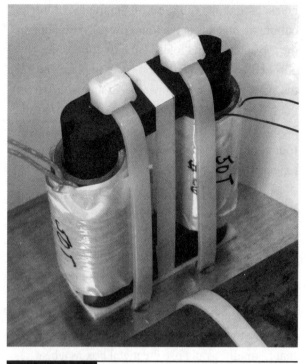

Figure 6-22 Assembled transformer.

two layers of Mylar tape for electrical insulation purposes (Fig. 6-21, step 3) before starting the wire winding down the coil (steps 4, 5, and 6).

After each coil is finished with the correct number of windings, it is good to wrap the entire coil once with Mylar tape to ensure that the windings stay tightly together in place and do not open up over time. Finally, assemble the transformer, where the halves of the ferrite core may be held together with tape or a tie wrap (Fig. 6-22).

4b. Chassis and Components Assembly: Chassis

Cut and bend a chassis from 1/16-inch aluminum sheet, as shown in Fig. 6-23 (or use an existing metal box with these minimum dimensions). At the inside corners of the bend lines, drill 3/16-inch holes to allow room for bending (it is easiest to drill these

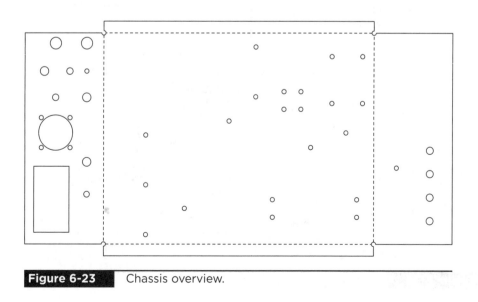

Figure 6-23 Chassis overview.

holes through the full sheet of aluminum before cutting the indents at the four corners. Also note that the top and bottom "folds" are only for cosmetic purposes to help hold the top cover in place and keep the sides from bending inward. As such, this part of the fabrication may be skipped, saving the time and labor of drilling the corner bend holes, cutting the four corners of the aluminum sheet, and bending the top and bottom folds.

Cut the holes for the component/board mounting (all chassis holes are 3/16 inch) and front and rear panels as shown in Figs. 6-24 and 6-25. Note that the dimensions probably will be different based on the dimensions of the components you use, so feel free to adjust these accordingly. If you use a wooden box, please note that extra heat sinks will need to be provided for the MOSFETs and other components that normally would use the metal box as a heat

sink; you also may want to include vent holes and a fan to circulate air because the circuitry will heat up more than normal if wood is used in place of metal (Fig. 6-26).

Cut an 8¾- × 6-inch (or whatever size you need to match your box) sheet of plastic to fit inside the front half of the chassis and insulate the Control Board and Line-Driver PCB from the metal chassis (see Fig. 6-29). Cut the holes to match those on the chassis.

Front-Panel Component Wiring

P1: red wire going to control board (see Fig. 6-6)

P3: black wire going to control board (see Fig. 6-6)

P4: white and black striped wire going to control board (see Fig. 6-6)

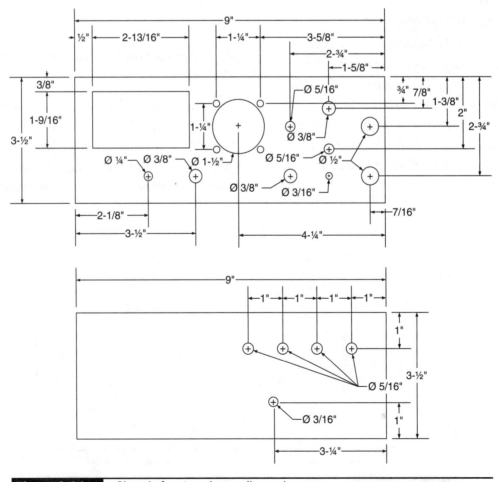

Figure 6-24 Chassis front and rear dimensions.

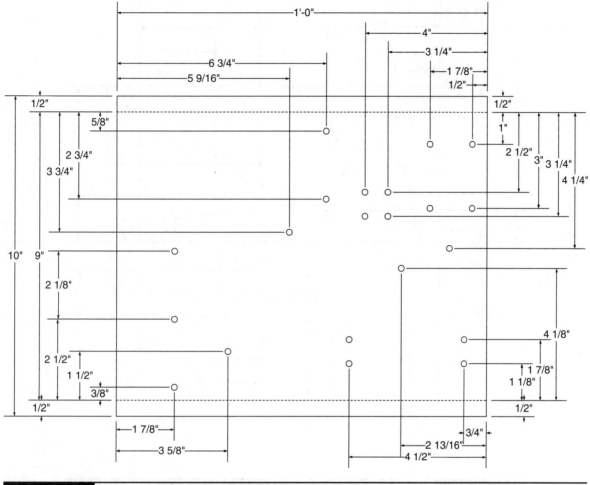

Figure 6-25   Chassis bottom dimensions.

P5: green wire going to control board (see Fig. 6-6)

P6: black wire going to control board (see Fig. 6-6)

P7: blue wire going to control board (see Fig. 6-6)

P8: white wire going to control board (see Fig. 6-6)

P9: yellow wire (center pole) going to control board (see Fig. 6-6)

P10: yellow wire (left pole) going to control board (see Fig. 6-6)

X2: green wire going to control board (see Fig. 6-6)

X3: blue wire going to control board (see Fig. 6-6)

X4: black wire going to control board (see Fig. 6-6)

FMFP: black wire going to control board (see Fig. 6-27)

NEON-R: red wire going to line-driver PCB (see Fig. 6-11)

NEON-B: black wire going to line-driver PCB (see Fig. 6-11)

R7R: red wire going to output/charger board (see Fig. 6-14)

R7B: black wire going to output/charger board (see Fig. 6-14)

M10FP: yellow wire going to output/charger board (see Fig. 6-14)

PWR-B: black wire from power cord to front-panel fuse F1A (see Fig. 6-31)

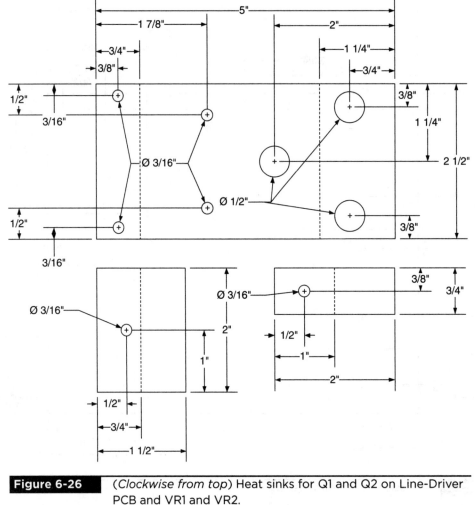

Figure 6-26 *(Clockwise from top)* Heat sinks for Q1 and Q2 on Line-Driver PCB and VR1 and VR2.

F1: black wire going from middle post of switch S1 to outer post of fuse F1A (see Figs. 6-28 and 6-32)

T2: black wire going from bottom post of switch S1 to chassis transformer T2CH (see Fig. 6-29)

S1-FP: black wire going from bottom post of switch S1 to line-driver PCB (see Fig. 6-11)

Figure 6-27 shows the front-panel components and wiring.

Rear-Panel Component Wiring

F1: black wire going from outer post of fuse F1A to middle post of switch S1 (see Fig. 6-27)

PWR-G: green wire from power cord to chassis screw ground (see Fig. 6-29)

PWR-W: white wire from power cord to wire splice and then going to line-driver PCB (see Fig. 6-11) and chassis transformer T2CH (see Fig. 6-29)

PWR: standard three-wire power cord plugs into 115-Vac wall power

Figure 6-28 shows an inside view of the Rear Panel.

Affix the front- and rear-panel components from Table 6-1 (parts list at the end of this chapter), as shown in Fig. 6-27 and 6-28. Mount the components and boards as shown in Fig. 6-29.

Figure 6-27   Front-panel components and wiring.

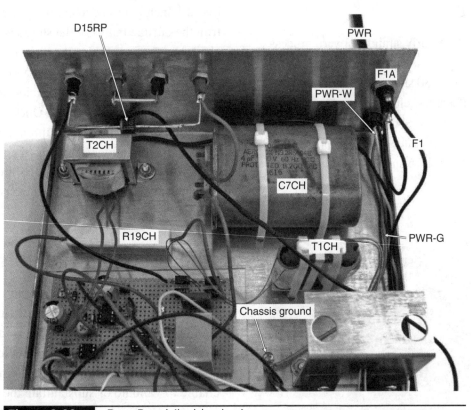

Figure 6-28   Rear Panel (inside view).

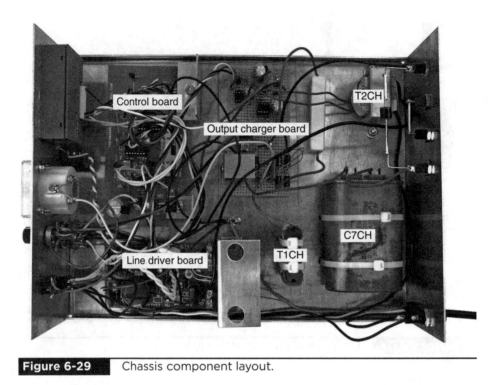

Figure 6-29 Chassis component layout.

4c. Chassis and Components Assembly: Coil

Coil assembly is a fairly straightforward process. Coils can be made in a wide variety of sizes and windings. Figure 6-30 shows a 50-turn coil (clear plastic is used here to help see the interworkings, but any color will do). Start with a 1-inch outside-diameter (OD) plastic tube sliced to approximately ¼ inch high with a 1/16-inch hole drilled for the

magnetic wire. Lexan sheets are then cut into disks (use a 1-inch hole saw to cut the center hole, and trim the outside radius of the sheets with hand clippers). Then superglue the disks to the outside radius of the tube (do one side at a time, let the glue dry before gluing the other side, and be careful not to touch the glue!). Once the glue has dried, feed the magnetic wire through the spindle hole (as shown in Fig. 6-30), and then wrap 50 turns. We then use a rubber dip (Hilltronix H-200) to coat the coils and help to hold the coil together (see Fig. 6-1).

Operating Instructions

Before employing this device for any particular ailment, you may want to consult with a doctor or qualified health practitioner who is familiar with alternative and complementary medicine. We do not claim to be doctors in medicine and are unable to validate or substantiate the claims for cures as stated or implied. We cannot assume any liability for claims, uses, and experimentation. We are presenting this information that we believe is real and authentic from qualified

Figure 6-30 Assembled coil.

researchers in these fields. The device is not FDA approved.

Because there are several controls and meters, you may find that applying decals (or tape labels) or marking with a fine indelible ink marker will help clarify the controls and their functions.

NOTE The basic coil can be used for just about all functions to some degree. But the effect *will* vary due to shape and size, and some designs are better suited than others to accomplish their intended purpose. You can build your own coil to whatever specifications you desire, or you can contact us and we can make one up that is best suited to your intended use. You may want to consult a professional or your personal physician in the use of pulsed magnetic therapy.

Familiarize yourself with the control descriptions and the more detailed explanations of controls and features that follow (Fig. 6-31).

S1 POWER SWITCH: Turns the unit on.

LED1: Lights up when S1 power switch is in ON position

S2 RANGE SWITCH: Position 1 selects "10 sec" and Position 2 selects "10 min" for sweep time of sleep and relaxation states. *This sleep-cycle function automatically causes the frequency to start at 100 Hz and drop down to 1 Hz over the selected period of time of 10 seconds or 10 minutes.* Positions 3 and 4 are for manual frequency adjustments: Position 3 selects a frequency range of 1 to 100 Hz, and Position 4 selects a frequency range of 50 to 1500 Hz.

R3 FREQ CONTROL: This is a 10-turn precision-control pot for accurately adjusting frequency ranges as selected by S2 Range Switch Position 3 or 4 and displayed on the Frequency Meter.

S3 CYCLE SWITCH: Selects this function to keep repeating or stopping after the first sweep of the sleep cycle. *The frequency change is read by the Frequency Meter.*

R2 GAUSS: Supplies voltage to energy cap producing joules = $\frac{1}{2}CV^2L$ of head = 50 μH, capacity = 4.4 μF, peak voltage = 450 V, peak current = 120 amps, turns on coil = 60, Gauss peak ≥ 6000+.

RELATIVE GAUSS: This meter reads the voltage on the energy-storage capacitor. This capacitor must charge to the selected voltage for the required energy to produce the magnetic Gauss. This charging function must occur during the period between pulses. The discharge is through a matching transformer and then to the terminals where the coil head is plugged in.

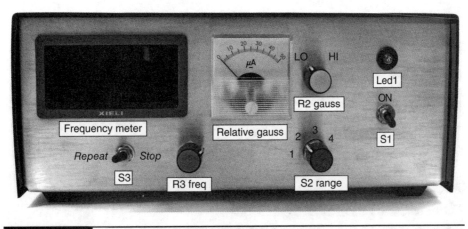

Figure 6-31   Front panel and controls.

There are two manually controlled frequency ranges on the unit set through the S2 range switch: Position 3 selects 1 to 100 Hz, whereas Position 4 selects 50 to 1500 Hz.

1. In Position 3 you will be able to adjust for maximum Gauss through the entire frequency range of 1 to 100 Hz.

2. In Position 4 you must adjust Gauss at one-half meter scale or the unit will overheat. Note that above 1500 Hz the Gauss meter will start to drop, but the unit is still functional at those higher frequencies.

You may experience some heating in the coil. If it gets hot, turn down R2 GAUSS control until the coil reaches a temperature that is comfortably warm.

Figure 6-32 shows the rear-panel inputs.

F1A: 1-amp fuse holder.

PWR: Three-conductor No. 18 power cord with green wire securely attached to chassis. Cord must be plugged in to a three-wire system for proper grounding and electrical safety.

J1-4: Banana jacks for hooking one to two coils on the rear panel. Single coil: Connect to J1 and J4, being sure to match colors. Two coils: Connect first coil to J1 and J2, and connect second coil to J3 and J4. Be sure to match colors.

Optional Coils

The device can be used with various coil sizes, from smaller coils for personal coverage, to a bed profile being placed between a mattress and a box spring, to covering a room of about 9 by 12 ft. There are many possible coil sizes and configurations that can be best matched to the target treatment area and desired effect.

CAUTION When using the device for head or neck applications, you must first consult a qualified practitioner or doctor.

Some Suggested/Common Uses of Coils

1. Standard 4-inch 6-kG flat Archimedean coil, unpotted. The is the coil we include in our sold units. A second of these coils can be used for sandwiching the target area or for another user.

2. Solenoid coils for wrists, hand, fingers, and erectile dysfunction and size enhancement.

3. Flexible flat 12-inch-diameter coil for sleep function (inserts under pillow).

4. Room-area coils up to 10 by 10 ft to cover an entire bedroom or to make a specialized treatment room for study of plants and so on.

5. Large solenoid from 2-inch-diameter and 4-ft-high coil.

Figure 6-32 Rear-panel inputs.

6. Separable coil for neck. *Caution: First consult a qualified physician.*

7. Coils for heads. *Caution: First consult a qualified physician.*

Test Procedure

- Check for 1-amp fuse.

- Do ac ground test: Turn power switch ON. Using Simpson meter, Set to RX10,000, connect the black lead to the ground prong on power cord, and touch the red lead to each blade prong. There should be no response on the meter.

- Turn all knobs full counterclockwise (FCCW). Power switch OFF (down). Sleep switch OFF (left).

- Plug paddle coil into the rear of unit (furthest opposing jacks).

- Put a soda can on paddle coil.

- Set all trimpots at 12:00.

- Set scope to 50 V/div at 10 µs/div.

- Connect scope probe to large blue resistor and negative rail on line-driver PCB.

- Connect LED/digital Simpson red lead to large-output capacitor and black lead to chassis (ground). Set meter to 200 V.

- Plug unit into ballasted mode.

- Turn unit on and set the line-driver PCB trimpot (502) to 40 µs.

- Turn the red and yellow knobs full clockwise (FCW).

- Turn the green knob to read 300 Hz on the digital panel meter.

- Adjust the trimpot (104) on the perfboard to read 100 V on the LED Simpson meter (amps should be 0.15 to 0.2 A; you should hear noise from the can).

- Set the panel voltmeter to 300 V.

- Move the power cord from ballasted mode to direct mode.

- Turn the green knob FCW, and adjust the 502 trimpot next to the 2206 chip to set panel meter to 1500 Hz (voltage on the LED Simpson should be 55 to 60 V; amps should be 0.24 to 0.25 A)

- Turn the red knob to Position 3, and turn the green knob FCCW; should read 2 Hz at 115 to 125 V.

- Turn the green knob to read 50 Hz; should read 110 to 120 V.

- Turn the red knob to Position 1 (FCCW "10 sec" mode).

- Connect scope probe to chassis ground and test point 6; set scope to 2 V/div at 1.00 s/div.

- Set trimpot (104) on PCB next to TP#6 to get 8- to 10-second cycle on scope;

- Move probe to TP#5; set scope to 2V/div at 2.5 s/div; should see a ramp signal

- Move probe to TP#2; should see a smaller ramp signal.

- Can vibration (noise) should slow down and go off for an instant and then come back on.

- Turn the red knob to Position 2 ("10 min"), and verify can noise.

Check Sleep Mode

- Switch the red knob to Position 1 ("10 sec" FCCW).

- Switch SLEEP MODE toggle switch ON (to the right). Can vibration (noise) should cycle once and shut down.

- Shut unit OFF, and let panel meter go to 0 V (about 20 seconds).

- Turn unit back ON; can vibration should cycle once and then shut down.

TABLE 6-1 Parts List

Ref. No.	Quantity	Description
		Control Board PCB
R1		1.2-MW ¼-W film resistor (BR, RED, GR)
R2		680-kΩ ¼-W film resistor (BL, GRY, YEL)
R5		1.5-kΩ ¼-W film resistor (BR, GR, RED)
R6, 15, 17		10-kΩ ¼-W film resistor (BR, BLK, OR)
R7		2.2-kΩ ¼-W film resistor (RED, RED, RED)
R8, 9, 10		1-MΩ ¼-W film resistor (BR, BLK, GR)
R11		220-Ω ¼-W film resistor (RED, RED, BR)
R14, 12		10-Ω ¼-W film resistor (BR, BLK, BLK)
R13		6.8-kΩ ¼-W film resistor (BL, GRAY, RED)
R16		1-kΩ ¼-W film resistor (BR, BLK, RED)
R18		470-kΩ ¼-W film resistor (YEL, PUR, YEL)
R19		5-kΩ horizontal trimpot
V2		100-kΩ horizontal MTD trimpot
C1, 7, 8	3	0.1-µF 50-V small blue polyester capacitor
C2		22-µF 35-V tantalum cap
C3		4.7-µF 25-V tantalum cap
C4		1-µF 25-V electrolytic vertical-mount capacitor
C5		10-µF 25-V electrolytic vertical-mount capacitor
C6, 9, 11		100-µF 25-V electrolytic vertical-mount capacitor
C10		10,000-µF 16-V electrolytic horizontal-mount capacitor
C12		2200-µF 35-V electrolytic vertical-mount capacitor
D1		1N4007 1-kV 1-amp standard-recovery diode
D2		1N4937 1-kV 1-amp fast-recovery diode
Z1, 2		6-V zener diode 1-W 1N4735A
Z3		4.7-V zener diode 1-W 1N5229
VR1		12-V regulator #7812 TO220
VR2		5-V regulator #7805 TO220
IC1		LM741 8-pin dip
IC2		LM324 14-pin dip
IC3		XR2206 function chip 16-pin dip
Q1, Q3		PN2222 *NPN* small signal gp transistor
Q2		PN2907 *PNP* small signal gp transistor
SOCK8X		8-pin IC socket for LM7418
SOCK14X		14-pin IC socket for LM324
SOCK16X		16-pin IC socket for XR2206
		Line-Driver PCB
R2, 3	2	5-kΩ 5-W power resistors
R6, 7	2	15-Ω ¼-W resistors (BR, GR, BLK)
R8		10-Ω 3-W MOX noninductive
R10		5-kΩ vertical trimpot
R11		1000-Ω ¼-W resistors (BR, BLK, RED)
R12	2	100-kΩ ¼-W resistors (BR, BLK, YEL)
R13, 14	2	47-kΩ 1-W resistor bleeders (not labeled on board) (YEL, PUR, OR)

TABLE 6-1 Parts List (*Continued*)		
Ref. No.	Quantity	Description
		Line-Driver PCB
RX		In-rush limiter 47D15 3-amps hot
C2		10-μF 25-V vertical electrolytic capacitor
C3		0.01-μF 50-V vertical film capacitor
C4		0.1-μF 400-V vertical met film capacitor
C5, 6	2	1.5-μF 250-V vertical met film capacitor
C7		0.0015-μF 630-V vertical met polyprop capacitor
C9		0.1-μF 50-V plastic capacitor
C10, 11	2	330-μF 250-V vertical electrolytic capacitor
C12		0.01-μF 2-kV disk capacitor
D1–4	4	1N5408 1-kV 3-amp SR diodes
D5		1N4937 1-kV 1-amp FR diodes
D7		1N4745 16-V zener diode ½ W
IC1		IR2153 high/lo side driver 8-pin dip
Q1, 2	2	500-V TO 247 MOSFETS IRFP450
PCLINE		PCLINEREV1
SOCK8X		8-pin IC socket for IR2153
SOCK247	2	3-pin connector mouser #158-P02ELK508V3-E for IRFP450s
		Charger Perf Board
RX1, X2		1-kΩ ¼-W resistors (BR, BLK, RED)
RX3		10-kΩ ¼-W resistor (BR, BLK, OR)
RX4		2.2-kΩ ¼-W resistor (RED, RED, RED)
RX5, X6		1-MΩ ¼-W resistors (BR, BLK, GRN)
RX8		100-kΩ trimpot
RX9		10-Ω 3-W resistor (BR, BLK, BLK)
RX10		1-MΩ 1-W resistor (BR, BLK, GRN)
RX13		10-Ω ¼-W resistor (BR, BLK, BLK)
CX1		0.001-μF 50-V disk ceramic cap
CX2, 3		1000-μF 25-V vertical electrol cap
CX4		0.47-μF 100-V MPP (474) cap
CX5		0.01-μF 50-V ceramic disk
DX1–7	7	1N4937 fast recovery 1-KV 6-amp rectifiers
DX8		5232 zener 5- TO 6-V zener diode
DX9, 10	2	1N5242 OR 1N4735 zener diode
DX11		1N5221B 2-V zener diode
DX12		1N914 60-V signal diode
DX13–16	4	1N4001 50-V 1-amp rectifier
SCR1CH		BTW691200 SCRs 50-AMP 1200-V TO-3P
IX1		8-pin dip comparator LM741
IX2		6-pin dip OPTO ISO diode to *NPN* 4N35
SOCK6X		6-pin IC socket for 4N35
SOCK8X		8-pin IC socket FOR 741
		Front Panel
M10FP		2- × 2-INCH 50-μA panel meter

TABLE 6-1 Parts List (*Continued*)

Ref. No.	Quantity	Description
		Front Panel
FREQMET-FP		Digital frequency counter
FH1FP		Fuse holder and 1-amp fuse
BU1FP		3/8-inch bushing for neon
BU2FP		Clamp bushing for power cord
NEONFP		Small neon indicator lamp power "ON"
S1FPA, B		DPST small 115-Vac 3-amp toggle switch
S2FPA, B, C		3-pole 4-position rotary switch
S3FP		DPST small 115-Vac 3-amp toggle switch
R7FP		500-kΩ 17-mm linear pot for pulse energy control
R11FP		10-kΩ trimpot for ranging meter M1
R21FP		10-turn 100-kΩ pot for frequency adjust
CO1FP		3-wire no. 18 power cord
KNOB1-3FP		Use knobs with pointers of your choice for RX2FP, RX7FP, S2FPA, B, C
		Chassis
C7CH		3.6- TO 4-µF 600- TO 1-kV motor PF capacitor
T1CH		12-V 0.3-amp with 115-Vac primary 60 Hz
T3CH		Switching XFR 50T#20 PRI, 50T#20 sec wound on 44130-P Core with 2-mil air gaps
HSINKCH		2.5- × 3- × 1.035-inch al heat-sink bracket for q1, 2, fabricated
PLATE69CH		Insert 6- × 9- × 1/32-inch plastic insulating plate for under power and control PCB
		Rear Panel
J1-4RP	4	Two red/two black banana jacks
C8RP		0.1-µF 250-V PP capacitor
R14RP		15-Ω 10-W power resistor
		Other Miscellaneous Parts
	2	Solder lug for chassis ground
	4	Rubber feet
	9	½-inch 6-32 screw
	3	¼-inch 6-32 screw
	1	¾-inch 6-32 screw
	4	½-inch no. 6 bunt tip screw
		1 ft of 3/8-inch double-sided tape
		6 inches of 1-inch doubled-sided tape
	6	3/16-inch wire tie
		Heat-shrink tube for leads from C8RP
		Making One Paddle
1 VC	2	3½-inch od 1-inch id 0.062-inch plastic
		5/16-inch 1-inch od 7/8-inch ID clear tube
		50T #18 magnet wire (roughly 30 ft)
	2	Banana plugs (1 red, 1 black)
		10 ft no. 16 wire (for two 5-ft leads twisted together)

CHAPTER 7

Trigger/Igniter

Overview

High-powered unit generates adjustable-repetition-rate and HV electrical pulses that can, for example, electrify an automobile while driving through hostile environments, protect gardens or other areas from animal damage, ignite all types of devices requiring a hot energetic spark, and so on. The unit will trigger Marx generators, energy-discharge dumps, spark gaps, and small jet engines of all types. See a video of the high-energy spark at www.amazing1.com.

This is the trigger device for the high-energy pulser (HEP100) used with the coilgun of Chap. 13.

Hazards

By itself, this device poses no hazards. However, once this device is attached to a capacitor bank, it can be dangerous or even potentially lethal depending on the power of the capacitor bank. It is designed to release a large amount of energy in a very short time, and the obvious risks associated with such an event must be kept in mind when working with this device. The operator is strongly encouraged to be familiar with high-voltage (HV) operating and safety procedures before using this trigger/igniter/shocker for its intended purpose. Eye protection should be worn when making, testing, and operating this device.

Difficulty

Moderately easy. Basic soldering and mechanical assembly skills required.

Tools

Basic wiring, soldering, hand tools, and fabrication equipment.

CAUTION If this system will be used for high-repetition discharges (equal to or greater than three events per second), for single-event high-energy discharges over 2000 J, or for unknown undamped loads, then a blast shield should be fabricated to cover the spark gap area. With sufficient energy, the tungsten potentially could fracture and send shards flying.

Description

This item is shown in modular form (Fig. 7-1). It can be used to trigger and ignite several of our products and other devices that hobbyists or researchers may wish to make. The heart of the device is two robust 3/8-inch-diameter by 2-inch-long solid tungsten electrodes that provide the gap where the main discharge occurs (note that cheaper tool steel can be used, albeit at reduced performance).

The unit, as shown, also may be triggered by a third electrode injecting an HV trigger pulse, or it may be used in a self-breakdown mode, where the spacing of the electrodes is carefully adjusted to break down at approximately the required discharge voltage. Such a setup is very useful when using current sources to charge capacitors and have them discharge when a desired voltage is reached. When used in this way, the ignition section can be omitted to simplify construction of the device.

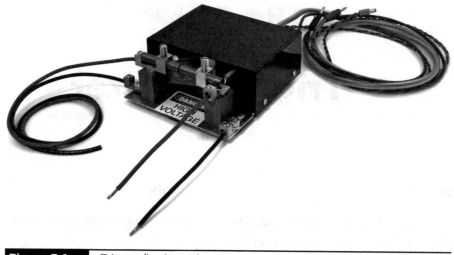

Figure 7-1   Trigger/igniter (shown without a safety shield).

We have used the gap in the self-breakdown mode discharging 5000 J as a single event, occurring about every 30 seconds. Obviously, we had a safety shield around the gap, made from heavy polycarbonate plastic, in case the tungsten rods shattered with the discharge. Care must be taken in the design of the discharge circuit when using this, or really any other similar device, in that the discharge curve is not so fast as to cause an explosion.

This is where knowledge of a pulse-forming network enters the equation and should be considered. The maximum possible energy released with this design is borderline to justify using a pulse-forming network, which is a complex system that requires specialized knowledge and advanced mathematics. High amounts of energy could be switched with a pulse-forming network, where the energy would be spread out over the time derivative. This would limit the very high peak energy that would take place without it. Pulse-forming networks require multiple capacitors and switches. Some mathematics is required, so for those of you unfamiliar with differential equations, we suggest that you search online and use some of the formulas and charts already presented (e.g., Wikipedia, etc.).

Construction of the TRIG100 is shown here with the trigger discharge circuit. This consists of a simple inverter transistor Q1 oscillating into a step-up transformer where the voltage output

charges capacitor C4. When capacitor C4 reaches the voltage breakover point of the SIDAC, its current discharges through it into the primary of T2. Because this is an HV pulse transformer, the secondary pulse goes through three 30,000-V isolation diodes properly biased to prevent any feedback of the main discharge back through T2. Other methods can be used to prevent this from happening, but we use this method.

In any system where capacitors and switches of this sort are being used, proper grounding is of paramount importance. **Ungrounded modules such as this during the experimental phase of setting up the system can kill you if they are not grounded and connected properly.**

The use of properly sized and insulated wire in the correct gauge is very important. In our lab, we use reasonably low-voltage jacketed high-current wire, but we add an additional sleeve over the wire appropriate for the voltage being used. These wires usually are not very long, and this is a convenient and low-cost method of effectively insulating the wire (use polypropylene, a stiff milky-white tubing found in most hardware stores).

The use of small toroidal cores in the charging leads not carrying the main discharge current is recommended because this will keep heavy and dangerous spikes from getting back to more sensitive circuitry elements.

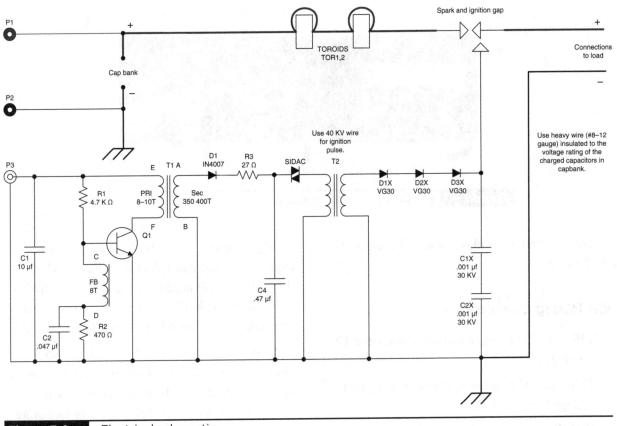

Figure 7-2 Electrical schematic.

Assembly

The schematic is shown in Fig. 7-2, and the component layout and wiring connections are shown on the printed circuit board (PCB) in Fig. 7-3. The PCB is available on our website (www.amazing1.com), or the unit also can be built on a standard perf board if the wiring layout

shown here is followed. This is not a complex circuit and should be fairly easy to build on a perf board.

Not all components of the PCB are employed because this board is used for other devices. Drill two holes through the board for nylon screws that will secure the board in place (Fig. 7-4; and the

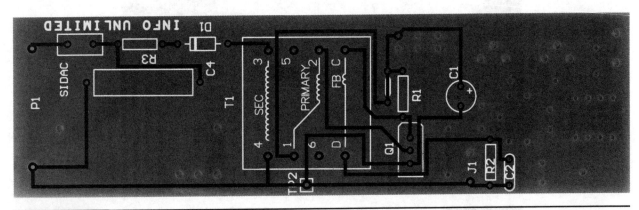

Figure 7-3 PCB and wiring connections.

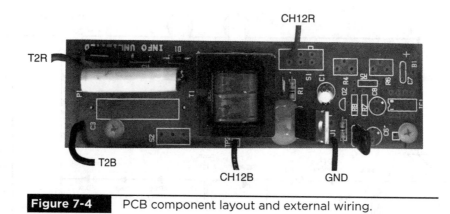

Figure 7-4   PCB component layout and external wiring.

two holes drilled in the plastic insulating sheet of Fig. 7-6).

PCB Wiring Connections

T2R: red wire going to chassis transformer T2 (see Fig. 7-11)

T2B: black wire going to chassis transformer T2 (see Fig. 7-11)

CH12R: red wire going to 12-V charger (see Fig. 7-11)

CH12B: black wire going to 12-V charger (see Fig. 7-11)

GND: black wire going to chassis ground (see Fig. 7-11)

Also note that the leads of the large yellow capacitor C4 are wider than the hole placed on the PCB and will need to be bent to accommodate these dimensions (Fig. 7-5). (This isn't necessary, of course, if you are using a standard perf board.)

Next, fabricate the chassis from 1/16-inch aluminum sheet about 6½ × 8 inches in size and the insulating sheet from 1/16-inch plastic about 5 × 4 inches in size. Cut or drill the holes for mounting components on the chassis and the charger wires through the back plate; the holes on the two side "arms" (bent upward in Fig. 7-6) should be tapped for 6-32 screws. Then bend a simple piece of 4- × 9½-inch plastic to cover and protect the circuitry, held in place with four 6-32 side screws, as seen

Figure 7-5   Modified wiring of C4.

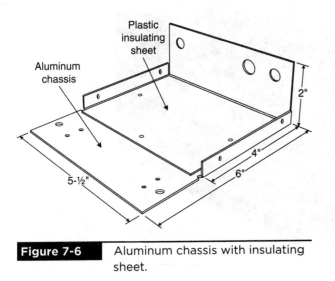

Figure 7-6 Aluminum chassis with insulating sheet.

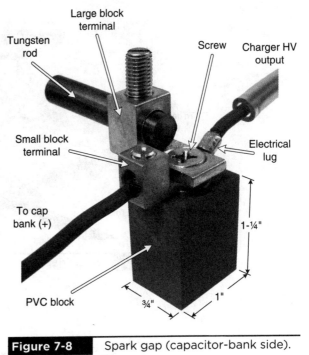

Figure 7-8 Spark gap (capacitor-bank side).

in Fig. 7-1. The holes drilled through the chassis do not need to be precisely positioned, but only to hold their respective components in place. Adhere some rubber feet beneath the chassis to keep it standing level.

The spark gap is held in place with two blocks (made of polyvinyl chloride or other structural plastic) about ¾ × 1 × 1¼ inches tall, secured with two screws from beneath the chassis and one screw on the top center to secure the electrical lug and two block terminals. The terminals attached to the device side have a slightly different configuration from the terminals attached to the capacitor bank (see Figs. 7-7 through 7-11).

Also, note that the device side of the spark gap has a clipped electrical lug beneath the block terminals acting as a spacer (Figs. 7-7 and 7-8), not connected to anything but there to keep the tungsten rod level with the other side (the capacitor-bank side) of the spark gap.

The short gap is held in place with a single block (made of polyvinyl chloride) about ½-inch square by 1-7/8 inches tall, secured with a single screw from beneath the chassis. Output from the T2 transformer goes through three diodes (D1X, D2X, and D3X) to an electrical lug on the short gap, through two capacitors (C1X and C2X), and then to the device negative (–) wire (see the electrical schematic of Fig. 7-2; also Figs. 7-9 and 7-11). Angle the brass ball on the short gap to within about 1/16 inch from the lug block (see Fig. 7-11).

The negative post connecting to the capacitor bank and negative post connecting to the device each consist of a block terminal and electrical lug secured to the chassis with a nut and bolt (Figs. 7-10 and 7-11).

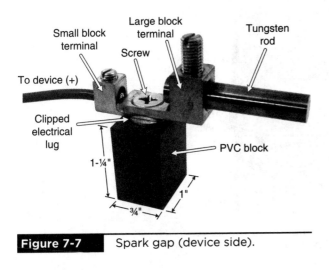

Figure 7-7 Spark gap (device side).

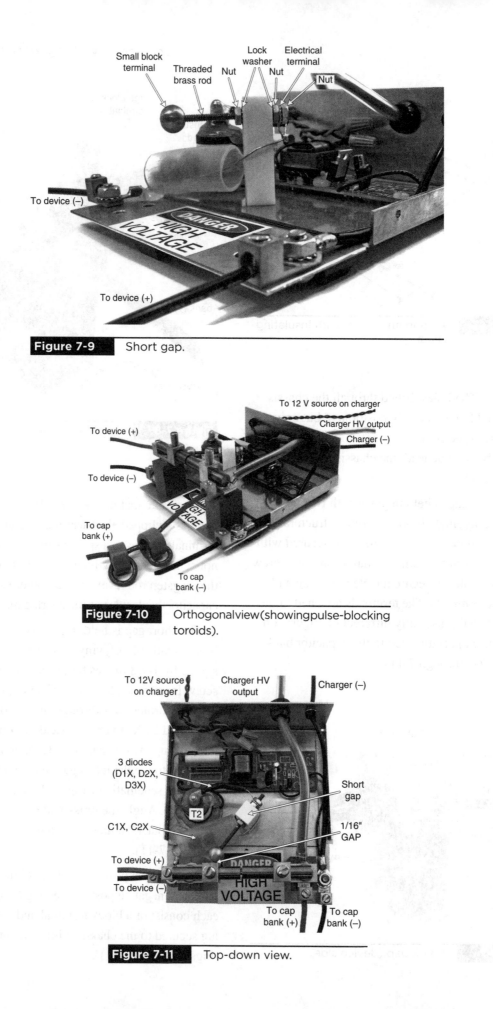

Figure 7-9 Short gap.

Figure 7-10 Orthogonal view (showing pulse-blocking toroids).

Figure 7-11 Top-down view.

Next, mount the components to the chassis (see Figs. 7-10 and 7-11). Transformer T2 can be held in place with a silicon adhesive/sealant; the top of this transformer and the exposed electrical connection also should be coated with silicon to minimize HV arcing or discharge.

Once the circuit board and all components are in place and mounted to the chassis, place the cover over the device to protect the components and reduce the chance of accidental discharge. The spark gap is left exposed so that the gap width may be easily adjusted to the proper threshold, but a safety shield should be constructed to cover the spark gap and protect against the tungsten rods exploding, which is a probability if any high-energy discharges will occur. The safety shield can be constructed to the user's needs (e.g., a front and top shield only will still allow easy access to adjust the spark gaps but will not stop shrapnel from exiting the sides and back). Some plastics offer better ballistic protection than others, and thicker shields will provide better protection than thinner shields. In addition, safety glasses should always be worn. A simple concept for a shield is shown in Fig. 7-12; a more complete and secure shield may be required.

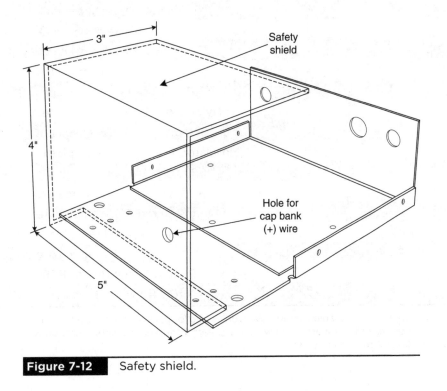

Figure 7-12   Safety shield.

Ref. No.	Quantity	Description	DB Part #
TABLE 7-1 Parts List*			
R1		4.7-kΩ ¼-W film resistor (YEL, PUR, RED)	
R2		470-Ω ¼-W film resistor (YEL, PUR, BRN)	
R3		27-Ω ¼-W film resistor (RED, PUR, BLK)	
C1		10-µF 50-V vertical electrolytic	
C2		0.0047-µF 50-V polyester film	
C4		0.47-µF 350-V polyester film	
D1		IN4007 1-amp 1000-V high-power diode	
SIDAC		300-V R3000	SIDAC
Q1		MJE3055 NPN	
T1		Small inverter transformer	TYPE1PC
T2		25-kV pulse transformer	CD25B
C1X, 2X	2	680p-0.001-µF 30-kV	680p/30kV
D1X, 2X, 3X	3	VG30 HV rectifier diode	VG30
	4	Small terminal block	
	2	Large terminal block	
	4	Terminal lug	
	2	PVC block 1¼ × 1 × ¾ inches	
		Nylon block 1-7/8 × ½ × ½ inches	
		Chassis 8- × 6½- × 1/16-inch aluminum	
		Insulator 4- × 5- × 1/16-inch plastic	
		Lid 4- × 9½- × 1/16 ft plastic	
		Wire nuts	
		Tungsten rod 3/8 × 2 inch pure	TUNG38
P1, 2		Banana plug 1 red, 1 black	
P3		2.5 mm-dc plug	
		Printed Circuit Board (PCB)	PCPSHK1
TOR1,2	2	Toroid trigger pulse blocker	TOROIDBLOCK

*Most parts should be available through electronics and/or hardware stores, although some may be more difficult to acquire, and these are listed with a "DB Part #" and are available through www.amazing1.com if they prove difficult to find elsewhere.

CHAPTER 8

Singing Arc

Overview

This device entertains and amazes. At first, people won't believe that sound is coming from a plasma arc, and even when they do, they will still have a hard time understanding how it works and how the arc can produce such clear sounds. This is a winning science project.

All figures are available in high-resolution full color as free downloads from www.amazing1 .com/eg3.

Hazards

Uses 12 V to power a small inverter that generates an inch of electrical plasma. Will cause a nasty and painful shock if contacted. Eye protection should be worn when making, testing, and operating this device.

Difficulty

Requires intermediate skills in wiring and soldering, along with experience using basic test equipment such as meters and a scope. Some basic sheet-metal work to form and fabricate the sheet-metal chassis, cover, and so on. You may obtain these with close enough dimensions from electronics supply houses such as Mouser or Digi-Key. Then all you must do is drill the holes. Electronic ability to use test equipment such as meters and a scope is necessary in completing this project.

Tools

Basic wiring, soldering, hand tools, fabrication equipment, voltmeter, and *low-cost oscilloscope*.

The singing arc is shown in Fig. 8-1, and the circuit schematic is shown in Fig. 8-2. The unit is based on one of our universal high-voltage (HV) modular power supplies. Circuit operates on 11 to 15 V direct current (dc), drawing 3 amps under full load, allowing portable battery or 115-V alternating current (ac) via a converter. Output voltage is a 60-kHz HV current that is fully short-circuit-protected. The high frequency also makes possible low-storage-energy voltage-multiplier stacks for HV dc sources as well as being an excellent plasma driver when used directly. Output current is fully

Figure 8-1 Assembled singing arc.

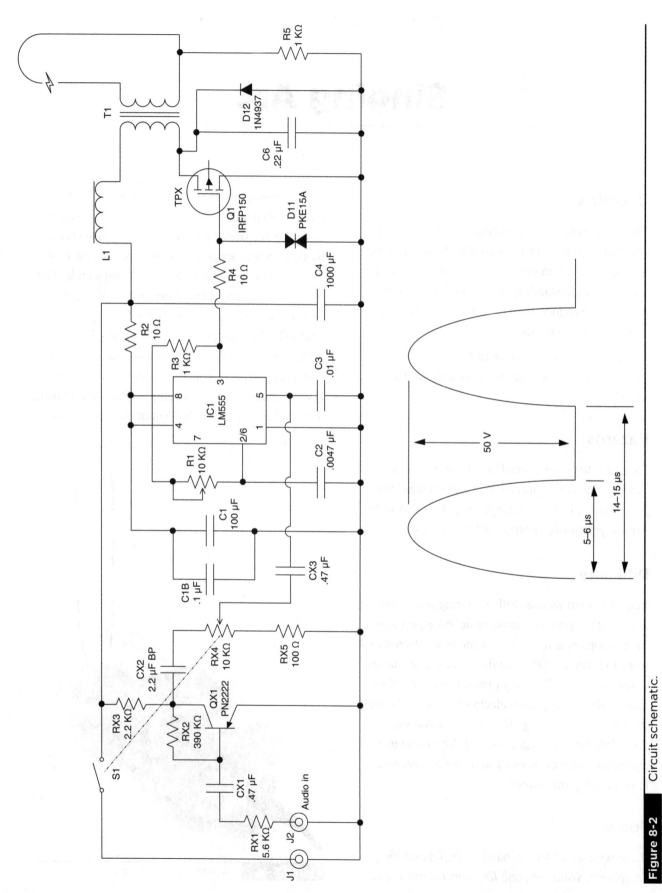

Figure 8-2 Circuit schematic.

adjustable via a control pot. The unit is excellent for powering neon and all types of gas-filled vessels using one or two electrodes, or can power objects simply by proximity. It easily retrofits to our voltage-multiplier modules that provide dc voltages up to 100 kV and currents of up to 0.3 milliamps. The current-limiting and control feature make this combination an excellent choice for charging capacitors for low-loss charging and 12-V portable operation. Also, it is an excellent choice for powering large and small antigravity craft, ozone air purification, and other applications requiring a HV current-controlled source. The module is shown built on a rugged printed circuit board (PCB) and is mounted in a plastic channel.

Specifications

Open-circuit voltage	7500 V peak at 60 kHz
Short-circuit current	10 milliamps short-circuit-protected
Input	11 to 15 Vdc at 3 amps fully loaded
Size	7 × 2-1/8 × 1-1/8 inches, weighing less than 5 oz

This is an intermediate-level project requiring basic electronic skills. Expect to spend $25 to $50. All parts are readily available, with specialized parts from Information Unlimited (www.amazing1 .com), and the parts are listed in Table 8-1.

The secondary of T1 is connected to a spark gap (also called a *plasma gap*; see Figs. 8-1, 8-2, and 8-13), where a stable, quiet high-frequency arc is formed. The modulation at this arc creates the sound.

Circuit Description

The primary of T1 is current-driven through inductor (L1) and switched at the desired frequency by FET switch Q1. Capacitor C6 is resonated with the primary of T1 and zero-voltage switches when the frequency is adjusted properly. *(This mode of operation is very similar to class E operation.)* The timing of the drive pulses to Q1 is critical to obtaining optimal operation.

The drive pulses are generated by a 555 timer circuit (IC1) connected as an astable multivibrator with repetition rate determined by the setting of the trimpot (R1) and fixed-valued timing capacitor (C2).

Power input is controlled by switch S1 that is part of control pot Rx4/S1. Actual power can be a small battery capable of supplying up to 2 amps or a 12-V, 2- to 3-amp regulated converter.

Theory of Operation

An audio-signal voltage from a radio, music player, amplified microphone, or other sound device is connected to audio jack J2, where it is fed to the base of input resistor Q1, where it is then amplified. The amplified signal is fed to pin 5 of timer I1, where it shifts the output frequency at pin 3 to the frequency of the audio signal; consequently, the arc frequency is now changing at the audio frequency, where one hears this as sound. The Rx4/S1 potentiometer controls the modulation shift. A current of 12 V is fed through jack J1 and is controlled by the S1 switch section of Rx4/S1.

Testing Steps

1. Preset trimpot R1 to midrange, and rotate Rx4/S1 to just click ON (lowest power setting). Short out the output leads using a short clip lead.

2. Obtain a 12-Vdc, 2- to 4-amp regulated power converter or a 12-V, 4-amp rechargable battery.

Note that this system can easily provide 30 W of usable power.

Assembly

The circuit is shown built using a PCB that may be purchased from www.amazing1.com for a nominal cost (about $10 at the time of this printing), requiring only that you identify the particular part and solder it into the respective holes as noted. Connections between components are simpler because the conductive metal traces are already on the underside of the board. Optionally, you may use a perf board, which is more challenging but often required for a science-fair project.

The perf-board approach is more challenging because now the component leads must be routed to each other (as per Fig. 8-10). We suggest that you closely follow the conductive metal traces on the PCB of Fig. 8-7 and mark the actual holes and circuit paths with a pen on the back of the perf board before inserting the parts so that everything will be laid out properly. When soldering, start from the lower left corner (if you're right-handed) as a reference, and proceed from left to right. *Note that the perf board is the preferred approach for science projects because the system looks more homemade.*

1. Lay out and identify all parts and pieces. Verify with the parts list; separate the resistors because they have a color code that shows their value. Colors are noted on the parts list. *If you are a beginner, we suggest that you obtain our GCAT1 General Construction Practices and Techniques, available as a free download at www.amazing1.com. This informative literature explains basic practices that are necessary in proper construction of electromechanical kits.*

2. The heat sink for FET Q1 is made from a bent piece of 1/16-inch aluminum sheet (Fig. 8-3). Two overlapping pieces should be used to ensure that the FET stays cool when operating. Note that the trailing edge is cut at an angle to provide clearance for some of the board components. Also note that you should drill the hole for attaching SW1/NU1 through the tab of Q1.

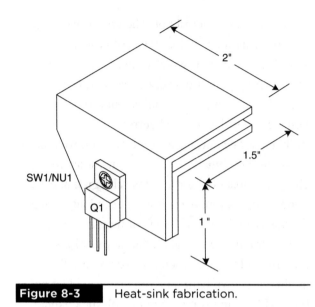

Figure 8-3   Heat-sink fabrication.

3. Assemble inductor L1 as shown on Fig. 8-4; it also can be purchased from our website (www .amazing1.com; Part # 6UH). To make this yourself, wind six turns of No. 18 magnetic wire around the bobbin. Use 0.015-inch air-gap shims on both sides of ferrite cores (as shown in Fig. 8-4); the easiest way to do this is with a strip of Mylar tape. Once this is in place, tape the entire assembly tightly together.

4. Fabricate the chassis from 1/16-inch plastic, a simple design that is basically a tray and a few

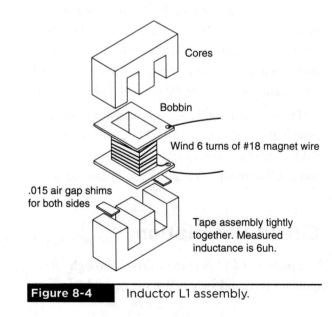

Figure 8-4   Inductor L1 assembly.

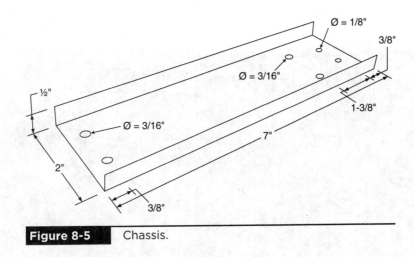

Figure 8-5   Chassis.

holes (Fig. 8-5). The hole positions don't need to be exact but only to match up to what is held in place on the chassis—the front panel, the PCB, and the insulating block.

5. Solder the components to the boards (see Figs. 8-6 through 8-10). If you don't have our PCB, then fabricate a piece of 0.1-inch-grid perf board to a size of 5 × 3 inches. Pay attention to the polarity of the capacitors based on their polarity signs and all semiconductors (Fig. 8-6; also the schematic in Fig. 8-2). Route leads of components as shown, and solder as you go, cutting away unused wires. Avoid wire bridges, shorts, and close proximity to other circuit components. If a wire bridge is

necessary, sleeve some insulation onto the lead to avoid any potential shorts.

When finished, double-check the accuracy of the wiring and the quality of the solder joints. Note the polarity of C1, C6, C18, and D12 and the orientation of I1.

Figure 8-7 is an "x-ray view" of the back of the PCB in order to show the wiring paths to the components as laid out in Fig. 8-6.

Note the capacitor C1B attached to the back of the circuit board, wired in parallel with capacitor C1 on the front. This is placed on the back of the board simply because of insufficient space on the

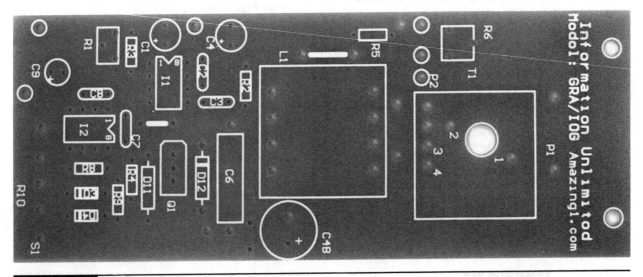

Figure 8-6   PCB parts identification.

C1B

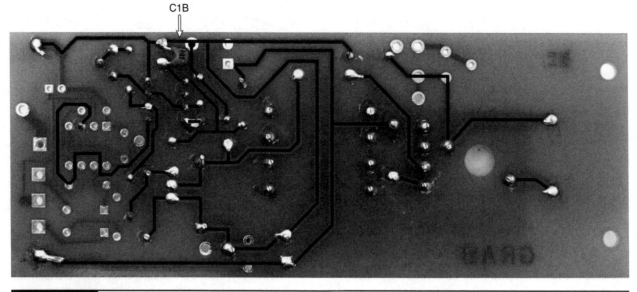

Figure 8-7   PCB foil traces ("x-ray view").

front, so if this circuit is being built on a standard perf board, the component layout can easily be arranged to fit capacitor C1B on the front.

> **NOTE** Before placing transformer T1 (the 28K089) on the board, apply a dab of silicone to the bottom of this transformer to help insulate it against the arcing that it is prone to do on this circuit. The larger hole cut into the PCB where the transformer is mounted is to allow excess silicone to push out and not obstruct the surrounding solder points. Alternatively, the transformer and surrounding circuitry may first be soldered in place so that the connections are established, and then the silicone can be inserted through the hole in the bottom of the PCB.

Note that Q1 is not shown soldered to the PCB in Fig. 8-8 because the heat sink would obscure some of the other components. Also note the jumper next to L1 that connects pin 3 of transformer T1 to capacitor C6 and diode D12 and the jumper next to I1 that connects pin 4 of the integrated circuit I1 to its own pin 8 and to the resistor R2.

PCB Wiring Connections

Wb: black wire going to Control Board (see Fig. 8-9)

Wy: yellow wire going to Control Board (see Fig. 8-9)

Figure 8-8   Components and external wiring.

S1: red wire going to Control Board (see Fig. 8-9)

HVR: high-voltage return wire going to one side of plasma rods (see Figs. 8-1 and 8-13)

P1: high-voltage output wire going to other side of plasma rods (see Figs. 8-1 and 8-13)

After all these components are soldered in place, the MOSFET Q1 with attached heat sink (which was fabricated in Fig. 8-3) is now soldered into the Q1 location on the PCB. The heat sink should clear the inductor L1 and transformer T1 by ½ inch or more.

Next, fabricate the control board as shown in Figs. 8-9 and 8-10 (there is no PCB for this subassembly). Note that Cx2 is a bipolar capacitor, so it has no polarity.

Control Board Wiring Connections

Ab: black wire going to front-panel audio jack (see Fig. 8-11)

Aw: blue wire going to front-panel audio jack (white wire in Fig. 8-11)

Pr: red wire going to front-panel power jack (see Fig. 8-11)

Pb: black wire going to front-panel power jack (see Fig. 8-11)

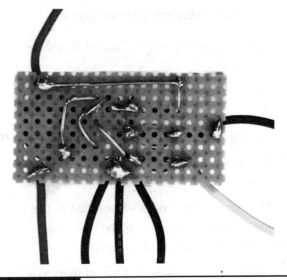

Figure 8-10 Control board wiring.

S1: red wire going to positive (+) 12-V rail on PCB (see Fig. 8-8)

Wb: black wire going to negative (–) rail of PCB (see Fig. 8-8)

Wy: yellow wire going to pin 5 of I1 on PCB (see Fig. 8-8)

Cut and drill the front panel from 1/16-inch aluminum sheet, and attach the components and wiring (Fig. 8-11).

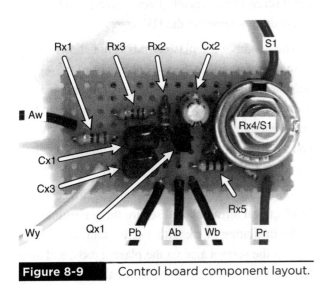

Figure 8-9 Control board component layout.

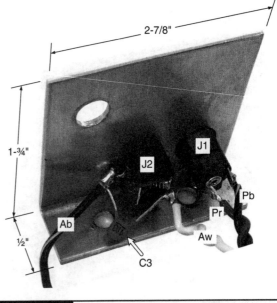

Figure 8-11 Front-panel components.

Front-Panel Wiring Connections

Ab: black wire going to Control Board (see Fig. 8-9)

Aw: white wire going to Control Board (blue wire in Fig. 8-9)

Pr: red wire going to Control Board (see Fig. 8-9)

Pb: black wire going to Control Board (see Fig. 8-9)

Then secure the dial Rx4/S1 of the Control Board through the top hole in the front panel (Fig. 8-12). Secure a knob to the dial.

Fabricate the arc rods, and mount them to a holding block (see Fig. 8-13). Note that the plasma rods are metal and get warm from the high voltage passing through them but will get hot where the plasma arc is generated. For this reason, the gap in the antenna should be kept at least a few inches away from anything flammable or meltable (e.g., the plastic casing if one is made for this device or the polyvinyl chloride block used to hold the base of the antenna).

The rod gap (where the plasma arc is created) can be orientated in any direction. A vertical gap may provide more consistency, and it may

Figure 8-12 Front-panel controls.

be possible to make it a little larger because the plasma will have a natural flow directly between the two points from bottom to top; however, the metal at the top of the antenna will heat up a good deal because it will be "bathed" directly in the plasma (so to reduce heat in the top rod, it is best to do a short run at the top of the plasma gap, as per Fig. 8-13, because a longer vertical run like that of Fig. 8-1 will get much hotter). A horizontal gap may "disconnect" more easily and may need to be a narrower gap but also will run cooler. Both designs (and all variations) will work, but it's up to you to decide which you like best or which is best suited to any design requirements you may have. Some good experiments can be conducted with different geometries and positioning of the plasma gap in addition to adjusting the gap length.

Fabricate an antenna mount from an insulating material, such as polyvinyl chloride (PVC) or wood, about $2 \times 1 \times 1$ inches (Fig. 8-13). Drill two holes into the top of this base, into which the metal rods will be placed (do not drill all the way through but about three-quarters of the block) and then two holes in the side of the base just deep enough to intersect the metal rods, which will be tapped and threaded for screws to contact the rods and deliver the HV output from the circuitry. When complete, attach the block to the mounting sheet next to the circuit board with an adhesive and/or securing it in place with a small screw from beneath (being careful that this screw does not contact the HV rods).

The height of the plasma rods is not shown to scale in Fig. 8-13; the overall height of these plasma rods is not crucial to the operation of this device, but they should be at least 4 inches high to allow enough of the heat to dissipate in the rods and avoid melting or burning the insulating block. Figure 8-1 gives an idea of a good overall scale.

Solder the HV wire and HV return wire each to an electrical terminal, and secure these terminals to the insulating block with a screw and nut. First, tighten the screws against the plasma rods, and then tighten the nuts against the electrical terminals.

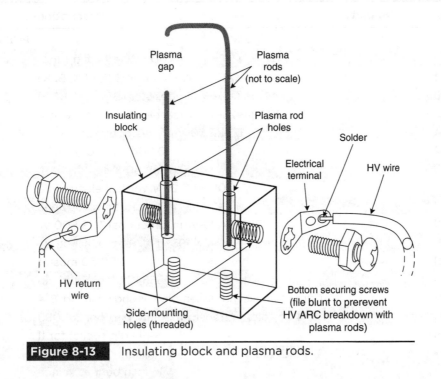

Figure 8-13 Insulating block and plasma rods.

(Components for the HV return wire are not shown in Fig. 8-13 for brevity.)

Once the insulating block assembly is complete, the three main parts are finished: the control board and front panel, the PCB, and the insulating block with plasma rods. Secure these three parts to the chassis with screws through the bottom; add some silicone to the area beneath transformer T1 for further insulation and a little more stability against the chassis.

When securing the PCB, first put a spacing nut between the PCB and the chassis and then another securing nut to hold the PCB in place (Fig. 8-14).

Then add four rubber feet to the bottom of the chassis (Fig. 8-15).

Secure the metal rods into the insulating block (if this hasn't been done already), plug in 12 V to the J1 power jack (usually a wall adapter), plug in an audio source to the J2 audio jack (such as a stereo or music player), turn the ON/VOLUME dial clockwise, and enjoy the sound made by the plasma of your Singing Arc!

Figure 8-14 Securing the PCB to the chassis.

Figure 8-15 Chassis bottom showing securing screws.

Ref. No.	Quantity	Description	Part #
TABLE 8-1 Parts List*			
		PCB Components	
R1		10-kΩ vertical trimpot	
R2, 4	2	10-Ω ¼-W film resistor (BR, BLK, BLK)	
R3, 5	2	1-kΩ ¼-W film resistor (BR, BLK, RED)	
C1		100-μF 25-V vertical electrolytic capacitor	
C1A		0.1-μF 50-V polyester film capacitor	
C2		0.0047-μF 50-V mylar	
C3		0.01-μF 50-V ceramic disk	
C4		1000-μF 25-V vertical electrolytic	
C6		0.22-μF metalized polypropylene	
D11		PKE15A 15-V unidirectional TVS 600 W	
D12		1N4937 1-kV 1-amp fast diode	
Q1		IRFP150 mosfet	
I1		LM555 IC timer 8-pin dip	
L1		Current choke info assembled as shown in Fig. 8-4	6UH
T1		High-voltage transformer info part no. 28K089	28K089
SOCK8		8-pin dip socket for I1	
Heat sink		2½- × 2-1/8- × 1/16-inch bent at 1 and 2½ × 2-1/8- × 1/16 inch bent at ¾ inch	
PCB		PCGRA8 PCB (or 1-7/8- × 5-inch perf board)	PCGRA8
		Perf Board Components	
RX1		5.6-kΩ ¼-W film resistor (GRN, BLK, RED)	
RX2		390-kΩ ¼-W film resistor (OR, WHT, YEL)	
RX3		2.2-kΩ ¼-W film resistor (RED, RED, RED)	
RX4/S1		10-kΩ 17-mm pot switch	
RX5		100-Ω ¼-W film resistor (BR, BLK, BR)	
CX1,3	2	0.47-μF 50-V metalized film	
CX2		2.2-μF 50-V bipolar vertical electrolytic capacitor	
QX1		PN2222 *NPN* general-purpose signal transistor	
Perf board		1¾ × 7/8 inch	
		Miscellaneous Parts	
Chassis		7 × 3 × 1/16 inch bent up at ½ inch on both sides	
Front panel		2¼ × 1-7/8 × 1/16 inch bent at ½ inch	
Knob		Red knob	
Power jack		2.5-MM DC jack	
Power supply		115-vac in, 12-vdc out 4-amp regulated	
Audio jack		3.5-mm mono jack	
Audio cord			
Feet			
Block		1- × 1¼- × ¾-inch PVC cube	
Electrodes		2 pieces of no. 12 buss wire	
Terminal lugs			
Screws			
Nuts			

*Most parts should be available through electronics and/or hardware stores, although some may be more difficult to acquire, and these are listed with a "Part #" and are available through www.amazing1.com if they prove difficult to find elsewhere.

CHAPTER 9

Tesla Lightning Generator (10- to 12-Foot Spark)

Overview

This large Tesla coil, standing about 6 ft high, is capable of throwing out sizable miniature bolts of lightning that when properly tuned will reach up to 10 to 12 ft in length (Figs. 9-1 and 9-2). The size, sight, and sound of these strikes are sure to amaze all onlookers, especially if they get inside a Faraday cage and watch the bolts of lightning strike to within inches of their faces!

Hazards

A serious, deadly electrocution hazard exists when using transformers with an output above 60 milliamps.

WARNING This is an advanced project for the experienced larger Tesla coil builder. These plans deal with and involve subject matter and the use of materials and substances that may be hazardous to health and life. Do not attempt to implement or use the information contained herein unless you are experienced and skilled with respect to such subject matter, materials, and substances. Neither the publisher nor the author makes any representations as to the accuracy of the information contained herein and disclaims any liability for damage or injuries, whether caused by or resulting from inaccuracies of the information, misinterpretations of the directions, misapplication of the information, or otherwise.

Eye protection should be worn when making, testing, and operating this device. Ear protection may be desirable because the operating noise levels can be quite high.

Figure 9-1 Early model experimental coil with 75 percent output power producing 8-ft discharges.

Figure 9-2 Two Tesla coils running in tandem.

Difficulty

Advanced. Considerable electronic and mechanical fabrication is required for this project.

Tools

Standard construction and electronic tools, fabrication equipment, 6-32 tap, and ¼-20 tap.

Special Note on These Plans

You do not have to follow the fabrication drawings exactly. They show how we did it but can be changed to suit your own availability of materials. The actual coil section is shown with photographs and hand-drawn sketches. The control and power section are a series of photographs that will show layout and assembly.

This very advanced electrical project is an excellent display for museums or can be a very fascinating and rewarding project for the serious and experienced hobbyist. Highly energetic audible and visual bolts of lightning jump into empty space, providing a spectacular effect. This can be an excellent advertising and attention-getting display when set up properly.

This project uses many basic materials but will require certain specialized parts. If you are unable to locate these components independently, they are available through our website at www.amazing1 .com.

Unit requires a dedicated 220-Vac 50-amp single-phase power source. It will require the same electrical plug that home arc welders use.

Safety is fully emphasized throughout these construction plans.

Basic Points and Safety to Consider before Building

Your Tesla coil produces large amounts of electromagnetic energy. It may damage computer systems and cause destructive interference to communication and sensitive electrical equipment. The coil (or equipment) should be in a shielded enclosure such as a Faraday cage if the coil is located near such equipment.

The primary circuitry consisting of transformers T1 to T6 (the power-supply box) produces lethal currents. Human contact with these points when system is connected to a power source ***will result in a fatal shock or serious burns.***

Never stand on a conductive surface such as cement or wet ground when operating this equipment.

Proper grounding of the system is very important for safe and optimal operation. This means that a separate ground wire must have a short and direct path to a dedicated, substantial ground rod.

Omission of the line bypass capacitors C2 and C3 shown in Fig. 9-4 located in the contactor box can create an unsafe condition with flashover in the primary feed lines.

Never operate this system in a flammable atmosphere because sparks can cause ignition. Low overhead wooden structures are also likely to be a fire hazard.

Always provide adequate ventilation because discharges produce large amounts of ozone.

It is often a merit of Tesla coil operation to make some kind of physical contact with the secondary spark discharge for demonstration purposes. ***This option should not be undertaken unless you have done this before with this power level and are absolutely certain of what you are doing.***

The secondary return of the output coil must be directly grounded to earth.

Never leave the system unattended where children or other unqualified personal may turn it on.

There is no need to energize the coil for longer than 10 to 20 seconds at a time.

This coil can produce discharges up to 12 ft in length (integrated over the entire path of the arc).

Therefore, we suggest that you position the main power switch at a remote point because sparks may jump to the operator if he or she is standing too close to the output coil. All metal controls should be insulated because contact may cause irritating burns to the fingers.

Do not use this system near pacemakers or other similar devices. Always warn spectators about the possible danger of being near this device if they are wearing or using sensitive equipment.

Do not operate this system near computers or other electronic equipment, especially those with solid-state and VLSI chipsets (including automobiles made since the 1980s owing to their computer-controlled systems).

A device of this type and power level should be operated from a dedicated 220-Vac 50-amp single-phase dedicated power circuit.

Brief Theory of Operation

Transformers (T1 to T6 in Fig. 9-17) step up household 220-Vac current to 15,000 Vac and charge the "tank" capacitor (C1 in Fig. 9-31). This capacitor now discharges through the four-gap spark switch (Fig. 9-31) on voltage peaks and steps a pulse of current into the primary coil (LP1 in Fig. 9-31). This sets up a resonant voltage of a frequency determined by the coil inductance and tank capacitor. Energy is now coupled into the secondary coil (LS1 in Fig. 9-31), also tuned to the resonant frequency of the primary circuit. The secondary energy now "rings down" with an exponential decaying waveform. The high-voltage (HV) output produced in the secondary is now a function of the ratios of LP1 primary Q to LS1 secondary Q factors or capacities. (It is important to note that voltage does not depend on the turns ratios but rather on the Q ratios.) The spark-gap switch must *turn off* so as not to allow the secondary "ring-down" energy to couple back into the primary circuit. The spark gap uses

multiple air-blown gaps to enhance positive turn-off and prevent ionization from excessive heating. The spark-gap electrodes should be tungsten for prolonged operation.

Circuit Description

The ***power-control schematic*** (see Fig. 9-17) shows transformers (T1 to T6) in a parallel combination to produce 15,000 V at 360-milliamp output. The secondary-coil midpoints are connected to a common neutral point to provide an ***earth ground***. This scheme provides 15 kV between the transformer end points but now has 7.5 kV with 720 milliamps from any end point to ground. The primaries are wired using standard 220-Vac wiring techniques. It is important to note the phasing dots shown adjacent to the primary windings.

> **CAUTION**   Note that the primary wiring must be isolated from secondary wiring.

Description of Major Components

Secondary Coil (see Fig. 9-38)

This is where the high voltage is produced. The coil form must be an excellent insulator and have a low dissipation factor to high-frequency currents. It preferably should be of a material that will not readily "carbon track" in the event of spark break-over and so on. Turns must be even and properly spaced. Turn crossovers or overlaps will always cause serious performance problems and must be avoided like the plague. The series resonant frequency of the secondary can be approximately calculated by considering it a quarter-wave section of length equal to the actual physical length of the wire used. A reduction in this figure can be fudged due to extra capacitance as a result of ionization at the top of the coil when discharge occurs. The voltage distribution along the length of the coil will be an R + sin/cos function, whereas current will be an R + cos/sin function, where R = radiation resistance, corona losses, output discharge, real power, and so on.

Output Terminal (see Fig. 9-38)

The output terminal of this system is shown with a 30-inch *toroidal* terminal. Such terminals can be expensive and hard to find, but their shape yields the best results (if needed, they can be found at www.amazing1.com). You may use stovepipe elbows as a substitute because sharp corners are not as degrading as they might be if used on a Van de Graaff generator. The purpose of the terminal is twofold. First, it electrostatically shields the top winding of the secondary coil from arcing into open air, which otherwise would cause burning of the coil and result in performance degradation. Second, the addition of electrical capacity to the top of a quarter-wave system enhances current flow through the coil. This property will increase spark energy at the cost of fewer discharges per unit time. Mathematically speaking, there is no limit to this capacity with the exception of the resonant frequency decreasing to an unworkable value. We are currently designing a computer program on this important property when used for voltage magnification and other non–magnetically coupled resonant systems.

LP1 Primary Coil (see Figs. 9-37 and 9-63)

This coil is combined with the tank capacitor and must form a series resonant circuit equal in frequency to that of the LS1 secondary coil with its associated output terminal and corona load (the *corona load* is the sparks coming from the output terminal; it is the real part of the notation for complex power when expressed in terms of $x + jy$). It is made tunable via a tap that allows connection anywhere along its spiraling turns. The material used here is 2.5- $\times$ 0.02-inch copper stripping. The tap can be two small pieces of copper sandwiching

the coil copper stripping and then clamped in place.

This coil and the connections must be heavy and secure to accommodate the high flowing primary tank currents as a result of its high Q factor. The tap settings are quite critical to operation and may take some time to set for maximum output. ***It is a good idea to have some assistance because you will be exposing yourself to deadly voltages.***

Coupling (see Fig. 9-60)

Secondary coil LS1 is coupled to the LP 1 primary coil and must be inherently tuned to the same frequency for efficient operation. Coupling of these circuits must not be too tight because beat frequencies may cause hot spots along the secondary coil. Too loose of a coupling, however, will not allow proper energy transfer between the circuits. These plans incorporate the results of many calculations and much experimentation to find the optimal placement between the primary and secondary coils that gives the best tuning such that adjustments need not be made (but if you are inclined, you may, of course, experiment by changing the position of LP1 by placing it on wooden blocks and so on).

SPKGAP Spark-Gap Switch (see Figs. 9-54 and 9-58)

This is where energy stored in the tank capacitor is switched into the primary inductor LP1. The spark-gap electrodes must allow for clean "makes" and "breaks." Adjustment is usually critical to allow C1 to charge sufficiently before breakdown or switching occurs.

Tuning on the high side of resonance will allow the gap switches to run cooler and smoother, and we find this best for reliable operation.

Remember that system energy is a function of the square of the charging voltage across the primary capacitor. It is important that the gap cleanly shuts down before the secondary current reaches it maximum value. The energy in the secondary must not couple back into the primary because it will cause erratic spark-gap operation, destructive voltage nodes, hot spots, and so on.

The spark switch used in this coil is four-point air-blown gaps using 3/8-inch *pure* tungsten for the spark-switch electrodes and is recommended if the coil is to be used frequently. Tool steel may be used because it is cheaper than pure tungsten, but it will degrade performance.

Tank Primary Capacitor C1 (see Figs. 9-43 and 9-58)

This is where the energy is stored that is exchanged with the primary inductor at a rate equal to the resonant frequency. It must be capable of handling and reversing high currents and have a low dissipation factor for efficient operation. Note that a special capacitor must be used in this circuit, and attention must be paid to its dissipation factor and reverse current-handling capability. Polypropylene is usually the dielectric of choice. A mounting bracket is shown. We chose a value of 0.05 µF, which is on the high side of the resonant peak by 15 percent or so. This allows full-voltage charging and smoother spark-gap performance along with cooler transformer operation.

Some builders are using series parallel combinations of CDE 942 series and GENTEQ 42 polypropylene capacitors, so this is another option to explore.

DC COX Suppression Circuit (see Figs. 9-12 and 9-15)

This part is necessary to block the high resonant frequency, spikes, transients, and harmonic voltages and currents from feeding back into the transformer. These currents can create destructive voltages that most certainly will cause premature breakdown of this part. Construction of this section is shown in detail.

Assembly

SPECIAL NOTE The following steps are those used by our laboratory when assembling this device. The builder may implement his or her own ideas, but we cannot guarantee the performance claimed if circuit parameters and values are changed. Dimensions shown also may vary with different mechanical parts. Make adjustments according to your needs.

Before you begin, study the plans, schematics, and photographs to get a general familiarity with how this Tesla coil will be built. Identify all parts and pieces that you need to purchase (see Parts List in Tables 9-1 through 9-3 at the end of this chapter).

IMPORTANT NOTE Hardware located near primary and secondary coils must be nonferrous brass; using steel or any ferrous metal near the coils will cause the metal to heat up to the point of melting plastic due to the oscillating magnetic field generated from the primary and secondary coils.

The general overview of the build starts with assembly of the Tesla coil itself, beginning at the bottom and working up, and then assembly of the power supply.

Assembly of the device is divided into three parts, each with subassemblies:

1. Power supply
   a. Contactor box
   b. Variac
2. Transformer bank
   a. Suppression circuit
   b. Transformer bank array
   c. HV bracket
   d. Safety spark gap (for transistor protection)
3. Tesla coil
   a. Primary coil
   b. Secondary coil
   c. Bottom base and legs

d. Safety gap

e. Capacitor bracket

f. Safety spark gap (for capacitor protection)

g. Base top plate

h. Coil wiring, adjusting tap, and breakdown baffle

1. Power Supply Assembly

The power supply (consisting of the contactor box and variac) takes an input of 220-Vac single-phase wall power and gives safe control from a distance, circuit protection, and with the optional variac, variable power output to the transformer bank (the transformer bank then converts this power to an output of 15 kVac at 360 milliamps for the Tesla coil). While the variac may be optional, it is highly recommended both to help tune the coil and for greater operational control.

The fabrication is a relatively simple mechanical and wiring assembly, which, along with the schematic shown in Fig. 9-4, should be enough for the experienced builder to complete. It is assumed that the builder is well acquainted with the electrical codes and requirements for a device of this power level. If not, then do not attempt this project without experienced help!

1a. Power Supply Assembly: Contactor Box

The input power to the Contactor Box is a 220-Vac 50-amp single-phase line (Fig. 9-3).

A suitable 30-amp 240-Vac disconnect with 30-amp fuses should be used for main power (Fig. 9-4). A simple push-button switch activating a 30-amp contactor can be used for operating, or just use the breaker switch itself. As mentioned, the system can be run through a variable transformer variac for output adjustments (not necessary but strongly recommended).

An electrical box (here with dimensions of $7 \times 11\frac{1}{2} \times 3\frac{1}{2}$ inches) may be used to house the

Figure 9-3 Main power plug 220-Vac 50-amp single phase.

components; see Table 9-1 for the parts list. Drill holes in the cover for mounting the switches, jacks, and indicator lights. And use a plastic insulating sheet or two (one to drill through and hold the BR84D bridge and 1000-μF 35-V capacitor in place and the other to insulate them from grounding to the electrical box; some dabs of silicone goop will hold the insulating sheets in place).

It may be unclear in Fig. 9-5 that the wiring of the contactor has HV terminals on its top and low-voltage terminals (with a diode soldered between them) on its bottom, so this is shown again for clarity in Fig. 9-6.

Given the relative shallow depth of the Contactor Box compared with the variac, a bracket may be made to equalize the heights (if the variac is used) and angled at a slight incline to make operating the controls a little easier (Fig. 9-7). This bracket can be made from either wood or bent aluminum sized to fit both the contactor box dimensions and variac height. The Variac Box and Contactor Box then can be mounted on a small wooden base (see Fig. 9-9) that can be put on the Transformer Bank Array (see Fig. 9-75) or on its own table farther back if more distance is required.

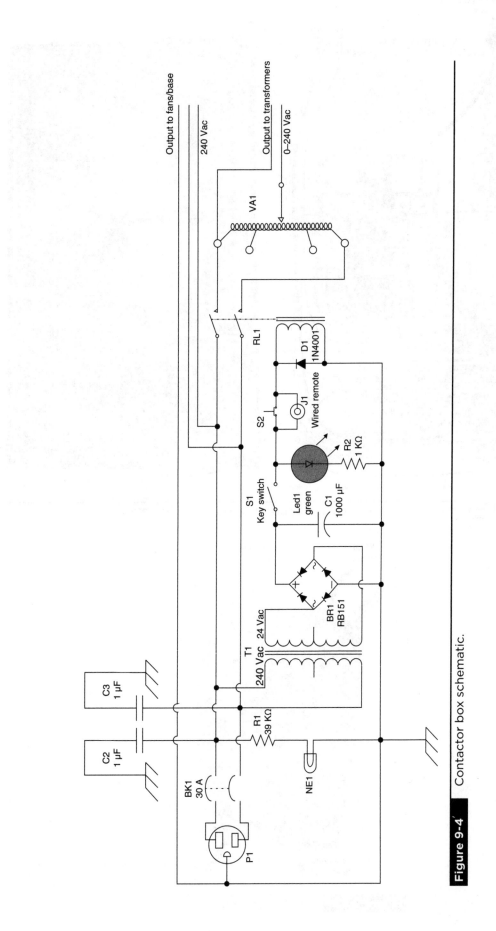

Figure 9-4   Contactor box schematic.

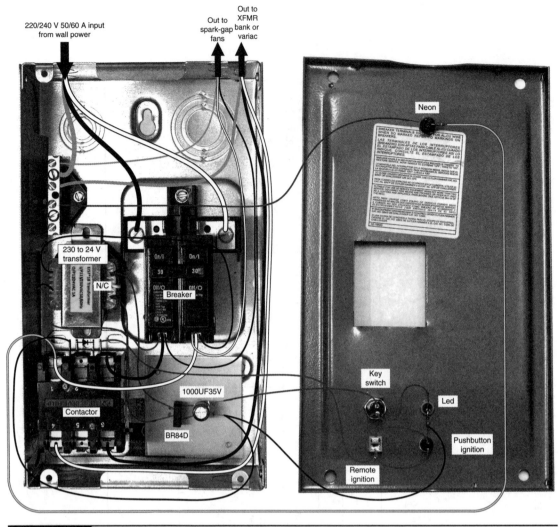

220/240 V 50/60 A input
from wall power

Out to
spark-gap
fans

Out to
XFMR
bank or
variac

Neon

230 to 24 V
transformer

N/C

Breaker

1000UF35V

BR84D

Contactor

Key
switch

Led

Pushbutton
ignition

Remote
ignition

Figure 9-5 Wiring of 100-amp contactor box (see Fig. 9-4 schematic).

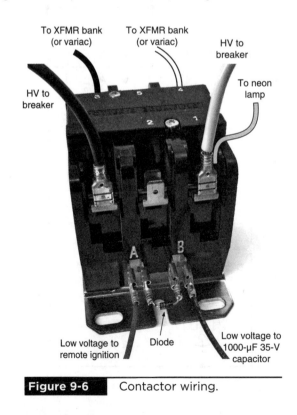

To XFMR bank
(or variac)

To XFMR bank
(or variac)

HV to
breaker

HV to
breaker

To neon
lamp

A

B

Low voltage to
remote ignition

Diode

Low voltage to
1000-µF 35-V
capacitor

Figure 9-6 Contactor wiring.

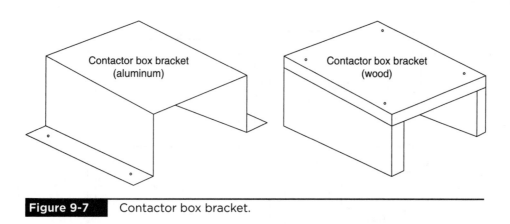

Figure 9-7   Contactor box bracket.

1b. Power Supply Assembly: Variac

The Variac (optional but highly recommended) is placed alongside the contactor box. An aluminum box to house the Variac can be made based on its size (Fig. 9-8 shows an example configuration; this should be sized based on the Variac, and remember that this box is optional and need only be made if the Variac will be used) with a plastic terminal cover over the back leads. Or you may find an already-made electrical box with dimensions large enough to accommodate the Variac. Or your Variac may come already enclosed in a box.

The terminal cover is made of plastic—and be sure to make two of these covers because one is used here for the Variac Box, and the other will be used to cover the HV output of the Transformer Bank (see Fig. 9-30).

Wire the Contactor Box to the Variac, if used (Fig. 9-10; see also Figs. 9-4 and 9-5).

The wiring of the Contactor Box to the Variac is fairly straightforward. Note that the locations of the input and output wires, in Fig. 9-10, going to the Contactor Box are different from those shown in Fig. 9-5, but the positioning of these wires doesn't matter as long as the internal wiring is done as per Fig. 9-5.

If desired, a simple remote ignition switch can be made to activate the Tesla coil from a greater distance. This is nothing more than a length of two-strand wire (cut to your desired length) with a

stereo plug on one end (that goes into the Remote Ignition jack on the Contactor Box front panel) and a push-button switch on the other end. Figure 9-11 shows this remote switch with a plastic tube and a couple of end caps on the switch side to make it a little easier to hold and operate.

Make sure that the remote ignition wire goes to the low-voltage control coil of the relay (the bottom pegs) and not to the HV upper terminals (see Figs. 9-5 and 9-6 for clarity).

2. Transformer Bank Assembly

The second part of the assembly involves the Transformer Bank Array that takes the feed from the power supply and converts it to an HV output of up to 15 kVac at 360 milliamps at the primary winding of the Tesla coil.

NOTE The power output from any single one of these 15-kVac 60-milliamp transformers is potentially lethal. Please exercise extreme caution when working on this transformer array, and triple-check to ensure that power is absolutely disconnected and the line is nowhere even near these transformers when they will be worked on.

2a. Transformer Bank Assembly: Suppression Circuit

You will note capacitors connected across each of the ac lines to the power common (ground)

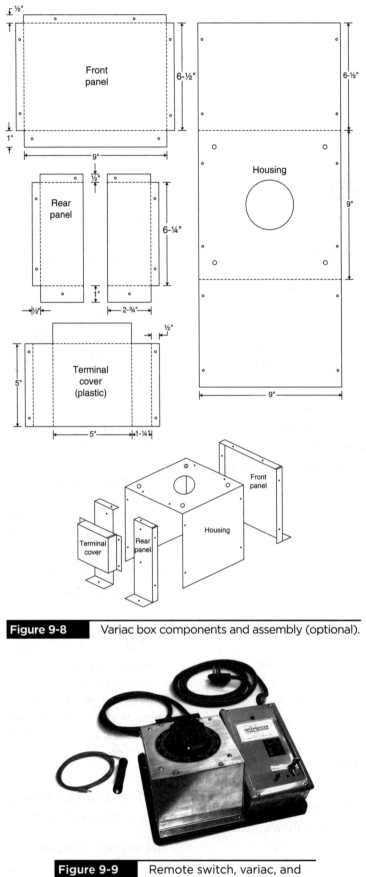

Figure 9-8 Variac box components and assembly (optional).

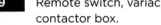

Figure 9-9 Remote switch, variac, and contactor box.

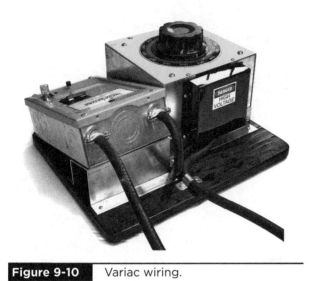

Figure 9-10   Variac wiring.

lead. These bypass any "kickback" pulses from entering the house wiring, where damage to sensitive electronic equipment is minimized (not eliminated!).

Suppression Circuit Wiring

9-kV unit: Use six 1800-V MOVs in series for each leg for MOV1, 2 P7215.

12-kV unit: Use seven 1800-V MOVs in series for each leg for MOV1, 2.

15-kV unit: Use eight 1800-V MOVs in series for each leg for MOV1, 2.

All units use four 300-Ω 50-W wire-wound resistors each for R1, 2, 3, 4.

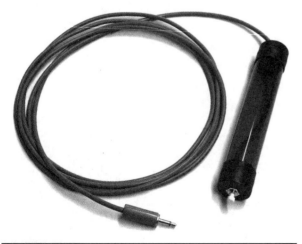

Figure 9-11   Remote switch.

All units use six pieces of 0.033-μF 1.6-kV MPP in each leg in series for C1, 2 P10501.

All units use coils of 80 turns of No. 24 polyvinyl chloride (PVC)–jacketed wire close wound on a 2-inch outside-diameter (OD) PVC tube for L1, 2.

Assemble the suppression circuit as shown in Figs. 9-12 through 9-16. Note that there are two polycarbonate boards: The top "mounting" board (see Fig. 9-13) is drilled to hold all the components and wiring that runs beneath the board. The bottom "insulating" sheet (see Fig. 9-14) prevents shorting of the circuitry and is attached to the top board with four screws through columns (we used ¾-inch-tall internally threaded joiner sleeves).

Solder components using smooth globular joints, and afterward, clean all flux to prevent flash-over (see Fig. 9-16). Internal green grounding wire is No. 16 to No. 18, which attaches to the No. 12 external earth ground. On the bottom view, note the soldering connection layout and holes tapped for component placement, how L1 and L2 are secured to the top plate by two screws each, and how the bottom sheet is smooth except for the four holes through which the spacer columns attach it to the top mounting plate.

When complete, this suppression circuit is secured to the transformer bank with flexible silicone adhesive (see Figs. 9-21 to 9-23).

2b. Transformer Bank Assembly: Transformer Bank Array

The 15-kV output is wired as shown in Fig. 9-17 and is connected through the (SUPP1) *suppression circuit* (see Fig. 9-12).

CAUTION   All grounding connections must be religiously secured to prevent a deadly shock hazard!

The Transformer Bank wiring is not inherently obvious, so Fig. 9-18 lays out the physical connections required (also see Figs. 9-21 and 9-23). Note to tap the

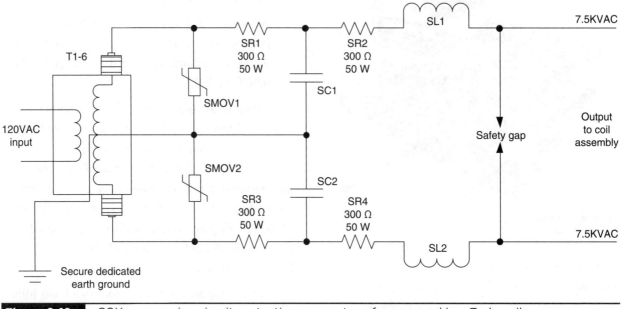

This circuit is property of Resonance Research of Baraboo, Wisconsin supplier of large museum quality Tesla Coils and Other Apparatus

| Figure 9-12 | COX suppression circuit protecting a neon transformer used in a Tesla coil.

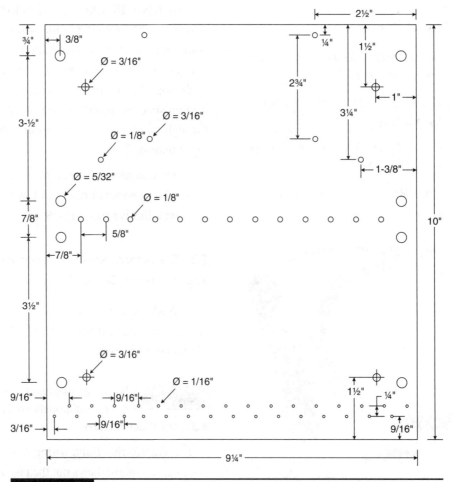

| Figure 9-13 | Suppressor layout board.

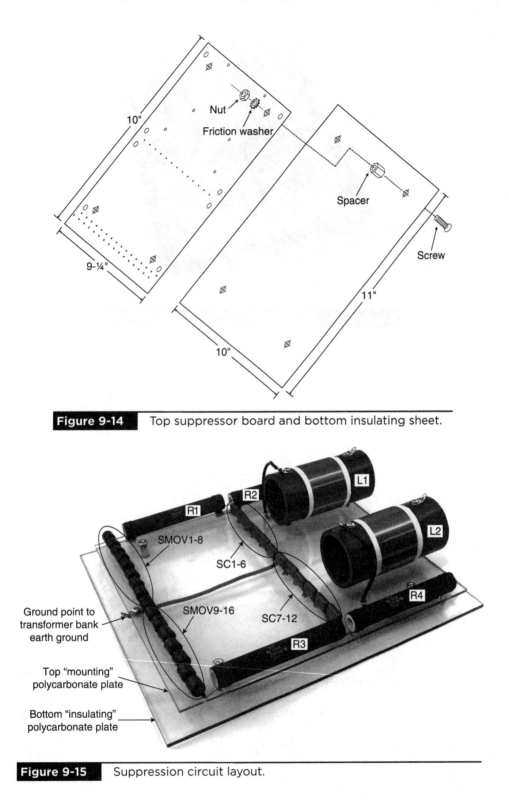

Figure 9-14 Top suppressor board and bottom insulating sheet.

Figure 9-15 Suppression circuit layout.

HV output off the central transformers to balance the load on the resistors.

Build a sturdy platform from plywood, with caster wheels, for holding the six 15-kV 60-milliamp transformers (Fig. 9-19).

Make two brackets from galvanized steel to hold the input and output wires of the power supply base (Fig. 9-20). The input bracket (on the "front" of the power supply base) has two holes for the fan input and variac input wires; the output bracket

Figure 9-16   Suppression circuit, bottom view.

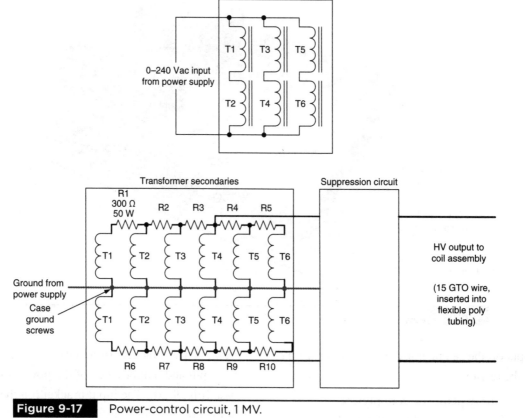

Figure 9-17   Power-control circuit, 1 MV.

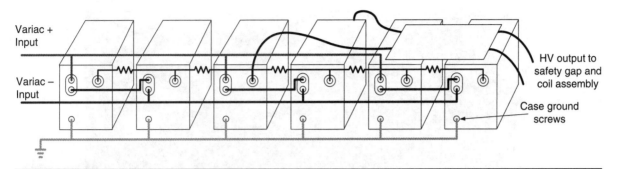

Variac +
Input

Variac –
Input

HV output to
safety gap and
coil assembly

Case ground
screws

Figure 9-18 Transformer bank phased wiring.

Figure 9-19 Transformers mounted to power supply base.

(on the "back" of the power supply base) has one hole for the fan output (the HV output is from a separate location; see Figs. 9-26 and 9-30). Put wire clamps on the holes to secure and protect the

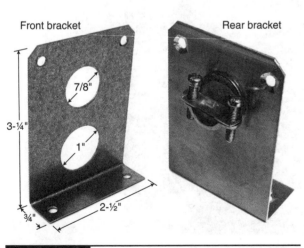

Front bracket

7/8"

3-¼"

1"

¾"

2-½"

Rear bracket

Figure 9-20 Wire brackets.

wires, and size the brackets and holes based on your needs.

Then wire up the transformers as per the schematics in Figs. 9-17 and 9-18, and connect the grounding strip and resistors (Figs. 9-21 to 9-23) and previously made suppressor circuit. Note that the primaries use No. 14 wire to interconnect.

Note the separating clearance of resistors away from the transformers to avoid electrical arc-over. And also note the metal grounding strap running along the base of the transformers (see Fig. 9-23).

2c. Transformer Bank Assembly: HV Bracket

Make two bracket legs from 1/16-inch aluminum, cut, drilled, and bent as shown in Fig. 9-24.

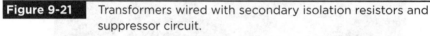

Figure 9-21 Transformers wired with secondary isolation resistors and suppressor circuit.

Figure 9-22 Transformer wiring (opposite side).

Figure 9-23 Wiring phasing detail.

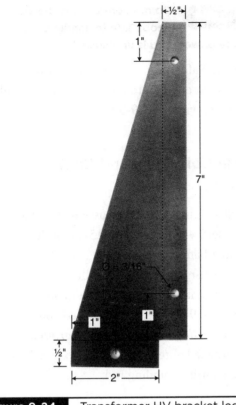

Figure 9-24   Transformer HV bracket leg.

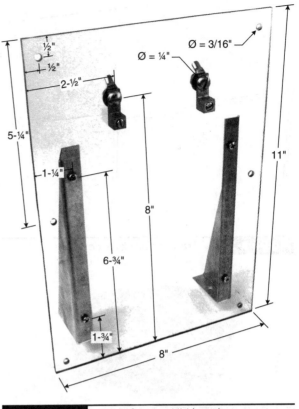

Figure 9-25   Transformer HV bracket.

Cut a 1/8-inch plastic/Lexan sheet with dimensions of 8 × 11 inches with holes drilled as shown in Fig. 9-25. The two holes for the electrical connectors/blocks are ¼ inch in diameter, with all other holes being 3/16 inch in diameter.

If the HV wires coming from L1 and L2 of the Suppression Circuit are too close to the metal transformer cases, they will have a parasitic loss through the grounded metal transformer cases. This is remedied by placing HV Transformer Bracket terminals high enough to keep the HV wires away from the grounded-metal transformer cases (Fig. 9-26). A plastic cover then is placed over these connections; refer to Fig. 9-8 for cover dimensions.

Once the Transformer Bank has been completed with the transformers, suppression circuit, HV bracket, and all the wiring, then fabricate a cover to enclose the base. Make this from plywood, with dimensions to match the base, and cut notches to accommodate the wiring brackets (Fig. 9-27).

Also drill four 1-inch diameter holes in each side of the box cover, and insert breathing plugs (Fig. 9-28) to allow some air circulation.

Figure 9-26   HV bracket mounted to power box and wired to suppression circuit.

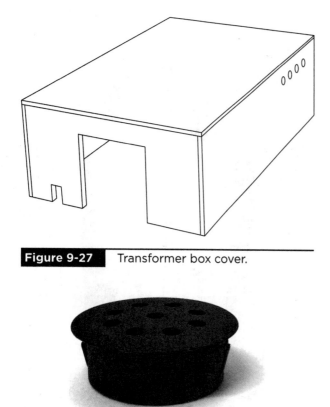

Figure 9-27 Transformer box cover.

Figure 9-28 Breathing plug.

2d. Transformer Bank Assembly: Safety Gap

The Safety Gap for overvolt protection has the same basic design as the Safety Gap in Fig. 9-49 but with an extra insulating layer beneath, as can be seen in Fig. 9-29.

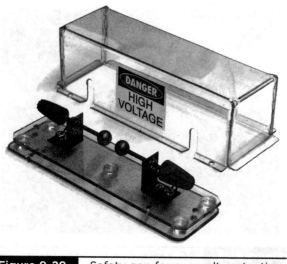

Figure 9-29 Safety gap for overvolt protection.

DANGER Do not consider the plastic knobs safe to handle or adjust unless the power cord is unplugged.

Once the Safety Gap has been built, wire it to the HV output and secure it to the box cover with screws as shown in Fig. 9-30.

3. Tesla Coil (Base and Coil) Assembly

The third and final part of the assembly involves the coil itself, which takes the 15-kV of the transformer bank and converts it into a lightning-like HV discharge of between 1 million to 1.5 million volts. The primary/secondary coil circuitry, as shown in Fig. 9-31, has a bottom lead connected directly to *earth ground and frame ground* via a No. 12 stranded wire lead and then externally to a dedicated-earth grounding rod.

CAUTION *Be certain that these connections are secure. This is extremely important!*

3a. Tesla Coil Assembly: Primary Coil

To make the Primary Coil LP1, cut a 2- × 2-ft base of 3/4-inch plywood (which will be used later as the base bottom) to temporarily mount the primary coil holding brackets and form the coil. Use a yardstick to draw two lines through the center in a cross-pattern at 90-degree angles to each other. Repeat again in an X pattern (these lines will be used to align the primary-coil bracket layout, as shown on Fig. 9-32). Then draw a 13-inch-diameter circle centered in the middle of the wood—one easy method is to tap a nail into the center of the wood and use a string and marker to measure out 6½ inches, which will give a nice 13-inch circle.

Now fabricate eight Primary Coil Holding Brackets from 1/16-inch clear Lexan (or other plastic if preferred), as shown in Fig. 9-33.

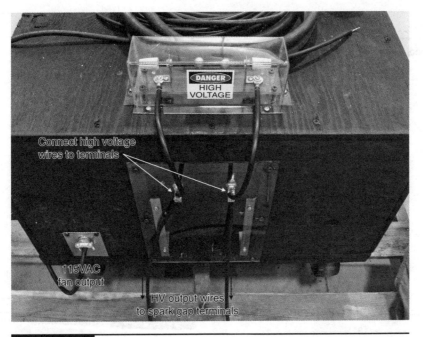

Figure 9-30   Power box HV wiring.

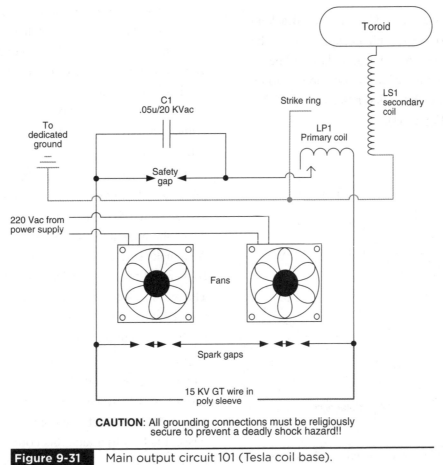

Figure 9-31   Main output circuit 101 (Tesla coil base).

Mark these lines

¾" plywood

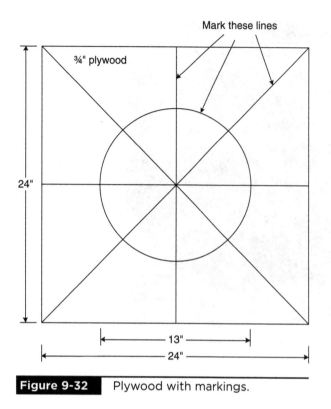

24"

13"

24"

Figure 9-32   Plywood with markings.

Secure the first Primary Coil Holding Bracket to the scrap piece of wood with a couple screws (these can be short wood screws because they will be removed later), with the bracket positioned along the first line and with the first notch on top of the 13-inch circle (Fig. 9-34).

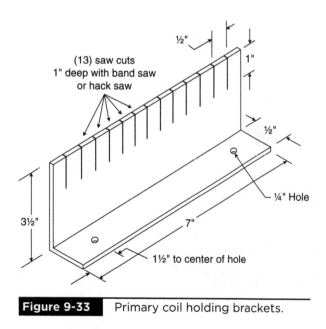

(13) saw cuts
1" deep with band saw
or hack saw

½"

1"

½"

¼" Hole

3½"

7"

1½" to center of hole

Figure 9-33   Primary coil holding brackets.

¾" plywood

First notch

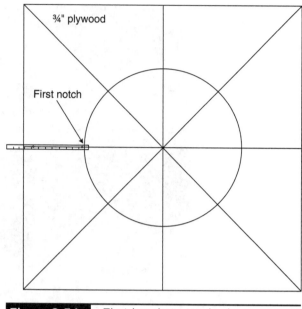

Figure 9-34   First bracket attached.

Repeat this for the seven remaining brackets on the seven remaining lines, going counterclockwise and moving each subsequent bracket another 1/16 inch away from the circle (so that the second bracket will be moved 1/16 inch away from the circle, the third bracket will be moved 2/16 inch away from the circle, and so on until the last bracket is mounted 7/16 inch away from the circle). This staggering outward of the brackets allows the primary coil to evenly expand ½ inch every turn (Fig. 9-35).

Before winding the copper ribbon to make the primary coil, solder a copper strip perpendicular to the length of copper ribbon such that it "drops" down about an inch away from the copper ribbon. Also make sure that this extra length of attached copper will be on the *inside* of the coil as it winds. The reason for this is to keep the terminal block and HV wire as far from the copper ribbon as possible to minimize arcing.

Punch a hole at the bottom of the copper strip, and attach a terminal block (Fig. 9-36). The block should be soldered and bolted to the ribbon to ensure the best electrical connection possible. Also, it is best to attach this copper strip and terminal block *before* the copper ribbon is wound

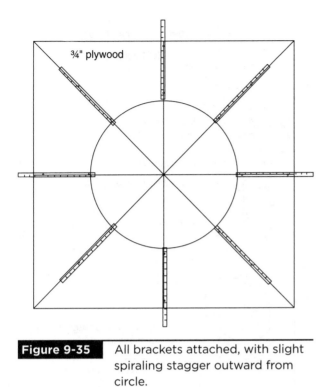

¾" plywood

Figure 9-35 All brackets attached, with slight spiraling stagger outward from circle.

into the primary coil because soldering it in place afterward can be difficult and may melt the bracket (see Fig. 9-72 for the final configuration).

With all the brackets secured in place on the plywood, wind the copper ribbon 11 times counterclockwise through the bracket notches starting about 1 inch before the first bracket (leaving enough overhang to attach the HV wire; see Fig. 9-72). When the eleventh turn is reached, spend some time making adjustments until the primary coil is nice and round; then clip the copper ribbon (see Figs. 9-63 and 9-70 for examples). Then use plumber's goop with a long and thin nose attachment to run a bead along opposite sides of each notch on each bracket (so that is roughly 90 notches total, or 180 lines of goop). Let the goop dry overnight. When finished, unscrew the brackets from the plywood, and remove the primary coil assembly. It will be structurally sound enough to hold together when picked up (Fig. 9-37). This may be set aside for now.

The reason for using this sheet of plywood as a base for making the primary coil instead of just building the primary coil straight onto the base top plate is so that the coil can later be freely positioned on the base top plate to ensure equal spacing around the secondary coil, as well as to minimize the amount of bracket overhang, resulting in a better-looking base unit (see Fig. 9-61 for the best geometry).

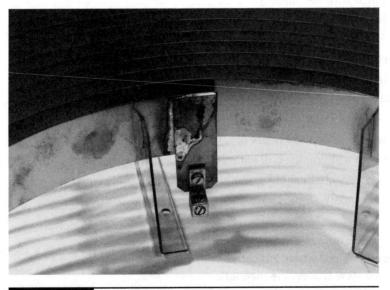

Figure 9-36 Primary coil ribbon attachment with terminal block.

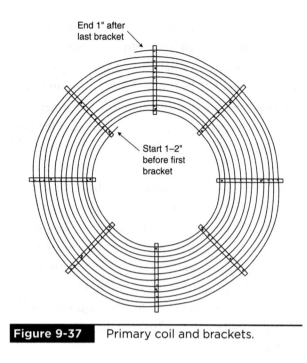

End 1" after last bracket

Start 1–2" before first bracket

Figure 9-37   Primary coil and brackets.

3b. Tesla Coil Assembly: Secondary Coil

Next, fabricate the secondary coil designated LS1 as shown in Fig. 9-38. The wiring around the PVC tube requires approximately 1020 turns closely aligned and straight and so will take some time, patience, and attention to detail. We suggest fabricating a lathe to hold the tube (similar to a wood lathe or spit roast) so that it can be spun to wind. It then can be rolled with one hand while the wire is guided with the other, or have a helper spin it, or, if you're more inclined, a small electric motor can be mounted at one end of the lathe to spin the tube. (See the QUASAR60 solid-state Tesla coil in Chap. 10, Figs. 10-84 to 10-91, for suggestions on how to build and use such a lathe.) This will take some time to make but will save many hours of frustration compared with winding the coil entirely by hand, which can be a very challenging assembly and prove difficult to get evenly aligned wire.

NOTE   Use brass fittings on the secondary coil where indicated. **Brass is nonmagnetic—if steel is used, the magnetic coil field generated will heat up the steel screws to the point where they will melt plastic.**

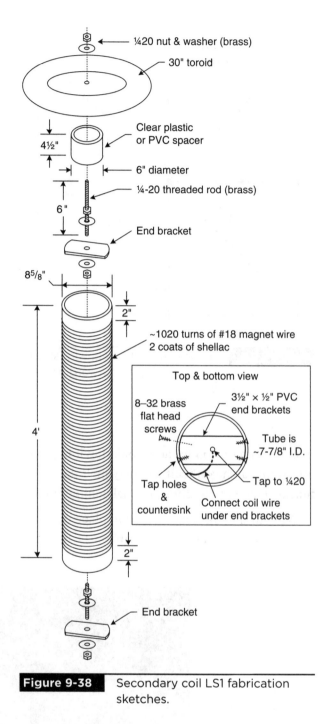

¼20 nut & washer (brass)

30" toroid

4½"

Clear plastic or PVC spacer

6" diameter

6"

¼-20 threaded rod (brass)

End bracket

8⁵/₈"

2"

~1020 turns of #18 magnet wire 2 coats of shellac

4'

Top & bottom view

8–32 brass flat head screws

3½" × ½" PVC end brackets

Tube is ~7-7/8" I.D.

Tap holes & countersink

Tap to ¼20

Connect coil wire under end brackets

2"

End bracket

Figure 9-38   Secondary coil LS1 fabrication sketches.

The end brackets are made from ½-inch-thick PVC sheet. PVC is preferred here because it is stronger, more malleable, and less brittle than other nonconducting materials.

Cut the flat PVC into a sheet 3 inches wide and at least 8 inches long. Locate, drill, and tap the center hole, and then use a bolt with a string and marker to scribe arcs with a 4-inch radius from the

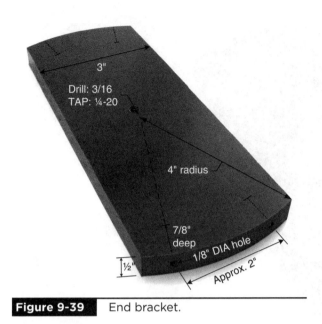

Figure 9-39   End bracket.

center-hole location. Then use a band saw to cut the PVC along the arc lines (Fig. 9-39). The bracket at this point should be a little longer than the inside diameter of the PVC tube for the secondary coil, so sand or grind down the bracket, making adjustments until it fits snugly and evenly into the tube. Then secure the bracket in the tube such that it is flush with the lip (use duct tape to hold it in place, if needed, but a good fit itself should be snug enough to hold the bracket in place). With the bracket in place, drilling through the tube and straight into the bracket will ensure that the holes are perfectly aligned. Drill the holes about 2 inches apart (this doesn't have to be exact) and about ¼ inch down from the edge of the tube (this part should be as exact as possible to keep the hole centered through the ½-inch-thick bracket). Drill through the PVC tube, and then continue another 7/8 inch into the bracket. Keeping the hole straight here is easiest done with a drill press; otherwise, just pay careful attention, and do your best by eye. Also, angling the drill toward the bracket's center point will keep the screws flush once they are in place against the outside of the secondary coil.

Fabricate two brackets, one each for the top and bottom of the secondary coil. These brackets

should be made before the secondary coil is wound because the winding is somewhat delicate when complete, so the rougher work of moving the coil around and drilling into it should be done first. If the brackets need to be removed for the tube to be wound (depending on how the coil lathe is built; see Ch. 10 Figs. 10-84 to 10-91), be sure that the brackets are marked relative to the tube so that they can be reinstalled with the correct orientation. A simple "I" marked across one bracket and tube lip and an "II" marked across the other bracket and tube lip will suffice.

When winding the coil, the start and end points of the winding should have three holes drilled through the PVC tube about ¼ inch apart, through which the magnetic wire will be threaded to hold it in place (Fig. 9-40).

The magnetic wire is then soldered to an electrical terminal (aluminum or brass), which is then secured to the PVC bracket with a brass threaded rod (Fig. 9-41). Do this at both the top and bottom, with minimal slack to avoid burning out the coil. Then secure the PVC bracket to the secondary coil with screws (Fig. 9-42).

After the secondary coil has been built, it is mounted to the base by securing the bottom bolt through the center hole of the base top plate, held

Figure 9-40   Threading of the magnetic wire.

Figure 9-41   Magnetic wire soldered to a terminal attached to a bracket.

Figure 9-42   End brackets and magnetic wire.

in place with a washer and nut from below (done later in assembly step 3g).

3c. Tesla Coil Assembly: Bottom Base and Legs

Using the ¾-inch plywood base that was used to make the primary coil, drill the four corner holes as shown Fig. 9-43 (but only drill the four corner holes for now because the remainder will be drilled later to match the dimensions and placement of the components as they are secured). Paint or stain the plywood as desired, and then set it on four casters—it would be best to make the bottom side the one that was marked up and drilled in step 1.

After the base is cut and the four corners are drilled, assemble the four base legs as shown in Fig. 9-44. It is okay to use a steel threaded rod instead of brass for these legs because they are far enough away from the primary and secondary to not heat up from the magnetic field oscillation.

Then secure the threaded rods through the base bottom with two nuts and two washers each (Fig. 9-45). Don't snug these up yet because you will probably need to adjust the "height" of the threaded rod such that it will extend the right amount past the base top plate, strike ring bracket, and washer while still having enough thread left to secure the top nut.

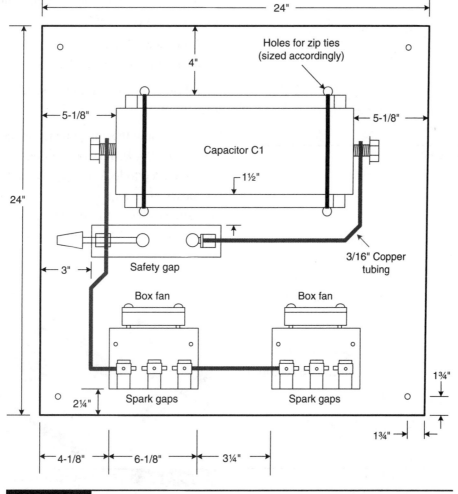

Figure 9-43   Base bottom.

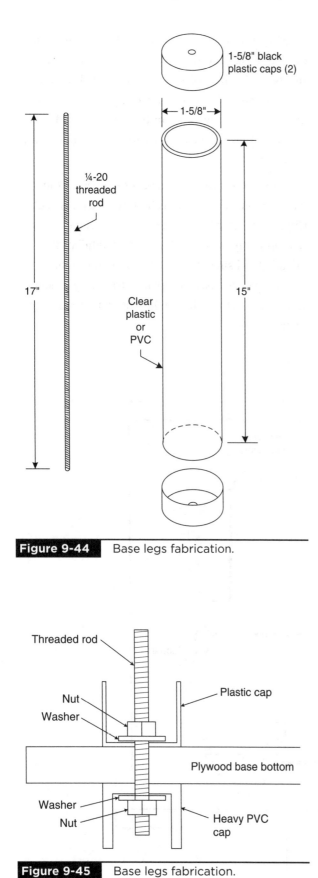

1-5/8" black plastic caps (2)

1-5/8"

¼-20 threaded rod

17"

15"

Clear plastic or PVC

Figure 9-44   Base legs fabrication.

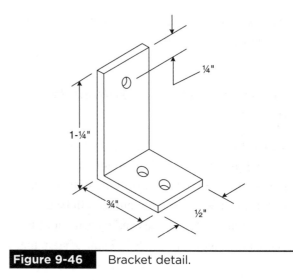

¼"

1-¼"

¾"

½"

Figure 9-46   Bracket detail.

3d. Tesla Coil Assembly: Safety Gap

Now fabricate the Safety Gap as per Figs. 9-46 to 9-49. Use 1/16-inch copper or aluminum bent into two brackets (Fig. 9-46).

Solder a 6-32 nut to the long arm of one bracket, and leave the other unsoldered (Fig. 9-47).

Fabricate a 1/8-inch Lexan sheet for the base (Fig. 9-48).

Then mount the components on the Lexan sheet (Fig. 9-49). The lock nut on the threaded rod is needed to lock the gap length but should be loosened up a quarter-turn or so prior to adjustments being made with the safety gap; otherwise, the friction will be too great, and

Threaded rod

Nut

Washer

Plastic cap

Plywood base bottom

Washer

Nut

Heavy PVC cap

Figure 9-45   Base legs fabrication.

Figure 9-47   Safety-gap hardware.

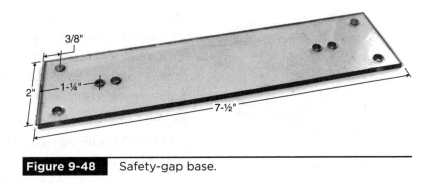

Figure 9-48 Safety-gap base.

turning the wire nut will merely strip the wire nut from the threaded rod. Steel rods and screws may be used here because this is a fault overload gap and not subject to constant use, and therefore, the steel components will not overheat as they would if used in or near an active circuit from magnetic oscillations (such as the secondary coil, the components of which must be brass or copper).

3e. Tesla Coil Assembly: Capacitor Bracket

Now make the capacitor mounting bracket (Fig. 9-50) from ¾-inch plywood using 1½-inch sheetrock screws (or similar) to hold the frame together; the bracket also may be secured to the base bottom, if desired, but this is generally not needed once the capacitor is held down with zip ties, which are run

through the holes drilled into the base bottom (see Figs. 9-43 and 9-58). We use 34-inch zip ties (they are about 1/16 inch thick by ¼ inch wide) to hold the capacitor C1 in place, which should be available at your local hardware store, or if you cannot find 34-inch zip ties, you can link three 13-inch zip ties or similar together to hold the capacitor snugly in place.

3f. Tesla Coil Assembly: Spark-Gap Switch

Next, fabricate parts for spark switch as shown in Figs. 9-51 through 9-55. Note that two spark-switch assemblies need to be made, and the hole locations on the mounting plates must be accurate, as shown in Fig. 9-51. Make the spark-gap mounting plates from 6¼- × 3½-inch sheets of 1/16-inch-thick plastic.

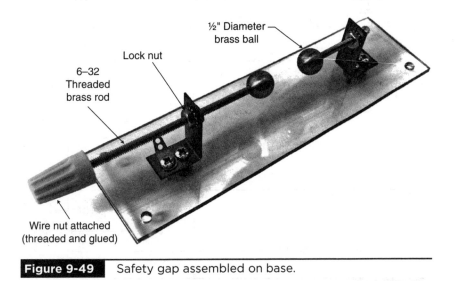

Figure 9-49 Safety gap assembled on base.

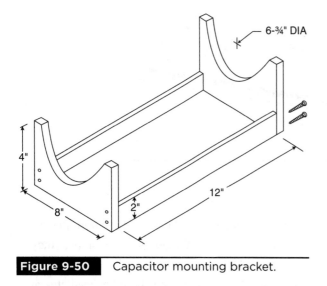

Figure 9-50   Capacitor mounting bracket.

The dimensions of the blocks and brackets are shown in Fig. 9-52, and the holes drilled in the PVC blocks should match the plastic mounting plates and galvanized steel brackets. Note that only two brackets require the notch for accommodating the front-facing electrical terminals at the far ends of either spark-switch assembly (see Fig. 9-54). Also note that these notched brackets must be bent in opposite orientations because one is used

for the left side and the other for the right side of the spark-switch assembly (again, see Fig. 9-54 for an example of how the notches work). The remaining four brackets require no notch and so are directionally interchangeable.

Assemble the hardware to the mounting plates as per Fig. 9-53, with the right spark gap having its terminal block on the right side and the left spark gap having its terminal block on the left side; the center "horizontal" electrical terminals should be connected by a 2-inch length of ¼-inch copper tube soldered fully into each spark gap (Fig. 9-54).

The protective spark-gap cover (bent from 1/16-inch black sheet plastic with overall dimensions of 5-7/8 × 16 inches) shields against the ultraviolet radiation produced by the discharge gaps and also directs more airflow through the spark gaps (Fig. 9-55).

Having assembled all the components, mount them to the Base Bottom as shown in Figs. 9-56 and 9-58.

Finally, attach a terminal block as shown in Fig. 9-57 with the input line coming from the standard 220-Vac "wall power" line (in this case,

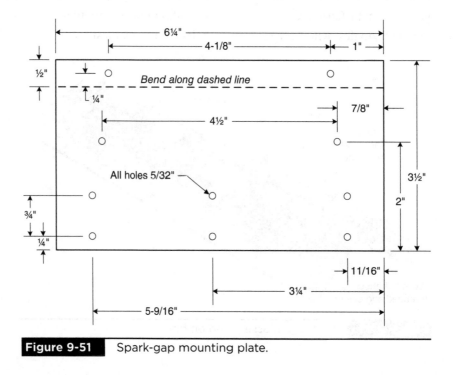

Figure 9-51   Spark-gap mounting plate.

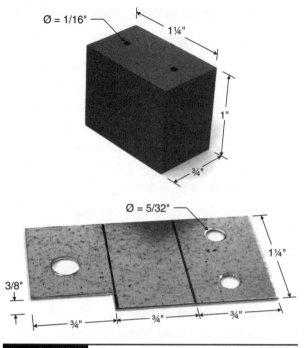

Figure 9-53 Spark-gap block and bracket.

Figure 9-52 Spark-gap block and bracket.

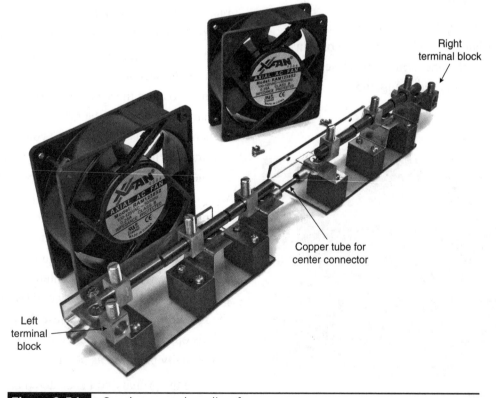

Figure 9-54 Spark gap and cooling fans.

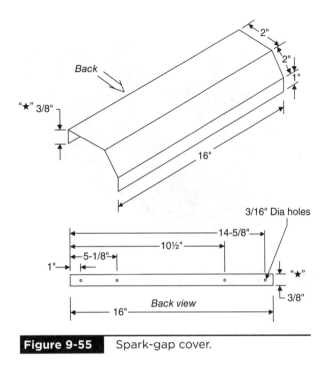

Figure 9-55   Spark-gap cover.

from the Transformer Bank Array; see Figs. 9-57, and 9-73), which is then used to run the 115-Vac spark-gap fans wired in series (due to 220-Vac line) and ground the strike ring.

When complete, the Base Bottom will have components laid out similar to those shown in Fig. 9-58 (note that the terminal block seen in Fig. 9-57 is not attached in Fig. 9-58).

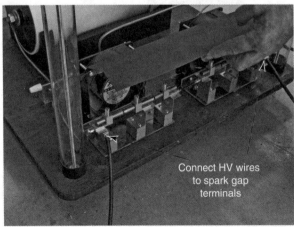

Connect HV wires to spark gap terminals

Figure 9-56   Spark-gap wiring.

3g. Tesla Coil Assembly: Base Top Plate

Next, fabricate the Base Top Plate as shown in Fig. 9-59. We use ¾-inch-thick plexiglass, but other material such as plywood also may be used (no metal, of course). Once this has been made, place it on the legs of the Base Bottom, where it can rest for now with the exposed threaded rods protruding through, because it does not yet need to be secured in place with end-cap nuts.

Now place the primary coil with brackets (previously made in Fig. 9-37) on the Base Top Plate in the orientation shown in Fig. 9-61 (it's rotated roughly one-sixteenth of a turn from the orientation in which it was made in order to minimize the amount of overhang and result in a better-looking base unit). The start of the primary coil winding, on the inner circle, should be located over the drilled hole, where the No. 8 wire will come through (again, see Fig. 9-61). The primary coil does not need to be precisely positioned right now, just roughly in place.

Then place the secondary coil on the Top Base Plate, with the bottom threaded rod going through the drilled hole (Fig. 9-60). This does not need to be secured in place with a nut from beneath because the secondary coil is just being used at the moment to align the spacing between it and the primary coil, which is the most important thing to do here. Spend as much time as you need to ensure that the spacing between the primary coil and the secondary coil is as equidistant as possible all the way around, with a combination of eyeballing the position and plenty of measurements. It is okay if the primary coil isn't perfectly centered on the Base Top Plate (which merely affects the appearance) as long as it is centered as precisely as possible around the secondary coil (which affects the function). Of course, because the primary coil spirals open, it can't be perfectly centered, but get it as equally balanced as possible.

Once the primary coil is centered properly around the secondary coil, use a felt marker

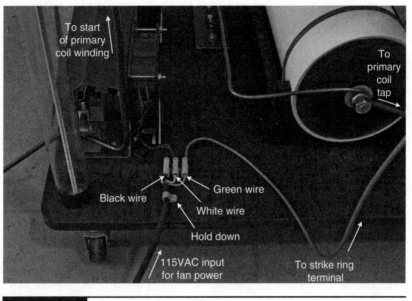

Figure 9-57 Base 220-Vac wiring.

through the bracket holes to indicate where they will be drilled on the Base Top Plate. Then remove the primary and secondary coils, and finally, either drill the marked locations straight through for the brackets to be held in place with nylon nuts and bolts (the easiest method) or tap the marked locations with a 1/16-inch drill bit for the brackets to be held in place with brass screws (looks better,

Figure 9-58 Base bottom component layout.

but more work and fussing around). Note that if the brackets will be screwed in place, be sure to use brass screws and not steel (or any other ferrous metal), which will cause all kinds of interference problems with the high-frequency HV current going through the primary coil. After the holes have been drilled (or tapped), secure the primary coil brackets in place on the Top Base Plate.

Next, fabricate the Strike-Ring Brackets (the corner positions of which are shown in Fig. 9-61). Make four of these Strike-Ring Brackets from 1/16-inch clear Lexan (or other plastic if preferred), as shown in Fig. 9-62.

The Strike Ring itself is made from ¼-inch-diameter copper tubing bent into a 32-inch-diameter circle. Normally, copper tubing is prone to kink when bent too sharply, but here the bend is gradual enough that kinking shouldn't be an issue. The tube can be bent by hand or wrapped around a cylinder of comparable diameter (such as a hot-water heater or car tire, for example) to get a rough size and then adjusted by hand.

Leave about a 3½-inch gap in the Strike Ring to keep it from becoming a shorted turn, which that is then near enough to the primary and secondary coils that

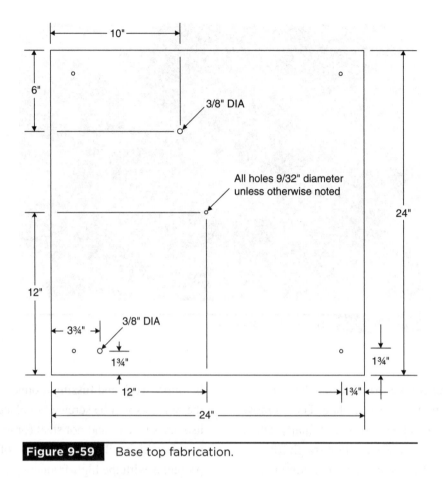

Figure 9-59 Base top fabrication.

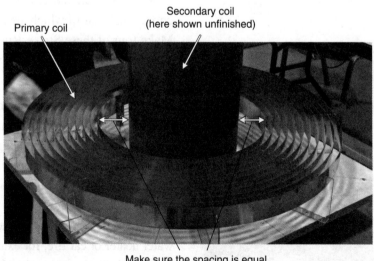

Figure 9-60 Centering the primary around the secondary.

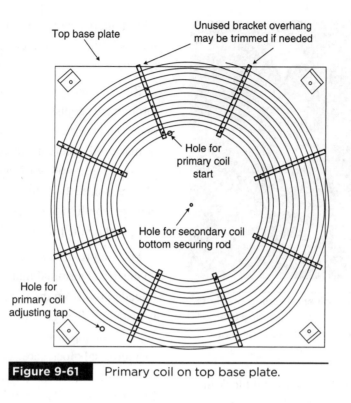

Figure 9-61   Primary coil on top base plate.

it would become an active part of this circuit and cause a parasitic voltage drain between the primary and secondary coils. The gap is there to prevent this. A piece of plastic tubing across the gap gives the Strike Ring the look of completion and also

prevents anything (such as clothing) from snagging the ring and pulling on it.

Now slip the Strike-Ring Brackets over the threaded rods of the legs of the base unit and secure them in place each with a washer and nut. Then attach the Strike Ring to the Strike-Ring Brackets with tie wraps (Fig. 9-63; note that the geometry of the primary coil brackets is not optimal in this image and should be rotated a little bit to achieve the orientation of Fig. 9-61 to reduce bracket overhang and give a better-looking result).

3h. Tesla Coil Assembly: Coil Wiring, Adjusting Tap, and Breakdown Baffle

Figure 9-64 illustrates how the base section components are wired to the Primary and Secondary Coils through the Base Plate Top. (Note that the three-lug connector is not yet attached here on the bottom board as it is in Fig. 9-57.)

Run one No. 8 wire from the spark gap's terminal block up to the bottom of the acrylic top plate, then around to the primary coil "start" hole, and then through the hole to attach to the start of

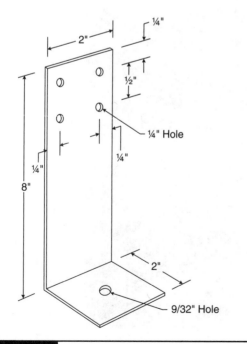

Figure 9-62   Strike-ring brackets.

Figure 9-63   Primary coil and holding brackets, strike ring, and strike-ring brackets mounted on the top base plate.

the primary coil (secured in the terminal block in Fig. 9-72). About 3 ft of wire will be needed for this, but cut it to fit properly. Also, run this wire through some ¼-inch inside-diameter (ID) flexible plastic tubing for further insulation, and then secure this along the bottom of the acrylic with plastic

wire holders (see Fig. 9-64; remember to use brass screws in the wire holders, not steel or anything ferrous).

Run another No. 8 wire from capacitor C1 through the "adjusting tap" hole in the top acrylic plate, which then runs around to the other side of the primary coil, where it attaches with the Adjusting Tap (see Figs. 9-64, 9-67, and 9-70). About 4 ft of wire is needed for this—the wire should have enough length so that it can wrap around to reach any point on the primary coil (Fig. 9-70).

For the ground line, one side is secured to an electrical terminal that is then bolted to the threaded rod from the secondary coil (once this is installed). The ground line then runs to the edge of the top base plate, where it is connected to a terminal block screwed to the bottom of the plate (Fig. 9-64). This then connects to a ground line (see Figs. 9-71 and 9-73) that should run to a dedicated earth ground rod or to a known substantial earth ground.

Make the Adjusting Tap by folding a piece of 4-inch copper ribbon in half to form a clip about

Figure 9-64   Primary and secondary coil wiring connections.

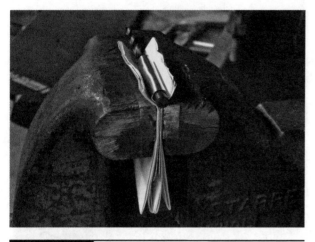

Figure 9-65   Forming the adjusting tap in a vice.

2 inches in overall length. Fold it around a cylinder a little bit larger in diameter than the No. 8 wire, such as a 3/16-inch drill bit, such that the No. 8 wire will fit somewhat snugly inside—not too tight of a diameter or the No. 8 wire won't fit into the tap, and not too loose or the soldered connection will be weak. Put the folded copper ribbon into a vice (paper the sides to prevent scuffing), and pinch it closed around the cylinder to bring the two arms of the ribbon next to each other (Fig. 9-65).

Some light tapping adjustments with a hammer may be helpful to make a more rounded circle (covering the copper with a scrap strip of plastic will help to more evenly distribute the blows and keep this part of the clip from looking too beat up). Ideally, the No. 8 wire will go into this clip with just the slightest bit of contact on all sides, and the solder will make good all-around contact between the No. 8 wire and the Adjusting Tap; feel free to

work the size of this opening until a good fit is made.

If the arms of the copper ribbon are a little bent, use the hammer and plastic strip on an anvil to straighten them out without scuffing them up (Fig. 9-66).

Once this part of the clip is built, cut the arms to equal length (if needed), and then use a needle-nose pliers to slightly bend the "lips" of the clip so that it will more easily slide onto the primary coil for later resonance adjustments (Fig. 9-67). Then solder the No. 8 wire into the clip.

Next, fabricate a "stick" to go around the clip, which will keep it tight and provide electrical insulation when the tap location needs to be adjusted on the primary coil. Make the stick from a 3- × 1- × ¾-inch wooden or PVC block; then drill a ¼-inch hole through the width of the block about 1¼ inch down its length. Then cut a line down the width to the hole, either with a band saw (if you have one) or put the block into a vise and cut it by hand, trying to keep the line relatively straight. You may need to cut a line two passes wide with a band saw (or use a wider-blade band saw, such as the sawing blade found on some multitools) to accommodate the copper ribbon clip without making it too tight. The block will not bend much, so spend some time making the cut wide enough for the arms of the copper clip to just fit snugly across the copper ribbon. Too tight and the clip is difficult or impossible to put on; too loose and the connection will be weak. Spend some time to ensure that a good connection is made.

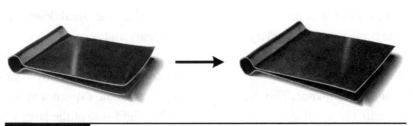

Figure 9-66   Straighten the adjusting tap arms.

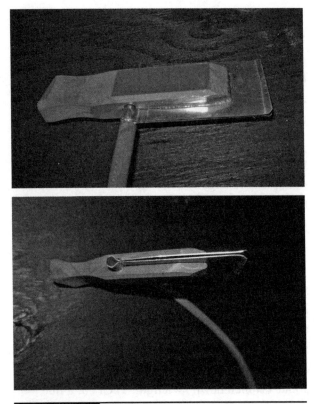

Figure 9-67   Adjusting tap.

Figure 9-68   Adjusting tap on the primary coil.

With the right tools, some edges and curvature can be cut to improve the appearance, but a straight block will work just fine, even if it may look a bit inelegant. Once the block is finished, slip it over the copper ribbon clip, and apply some epoxy to hold it in place. Now the Adjusting Tap is complete.

The Adjusting Tap then can be moved around the primary coil until the best position is found for a resonance that will produce the longest arcs (Fig. 9-68). There should be enough No. 8 wire such that the tap can reach any position on the primary.

Next, make a Breakdown Baffle to prevent breakdown between the primary and secondary coils, which will otherwise erode the insulation of the secondary coil wiring and impair the Tesla coil's operation and performance. This is a simple fabrication consisting of a rolled sheet of plastic held together with nylon nuts and bolts.

One shroud may suffice, but we feel safer with two, and the cost and effort are worthwhile for what is being protected. A single shroud with two layers can be made, but we think that two separate shrouds of different heights yields a better aesthetic effect. The shrouds don't need to have a precise diameter, but they must, of course, clear the secondary coil to fit around it. An extra 1 inch beyond the secondary coil's outer diameter is a good value, so if your secondary coil has an 8¾-inch outer diameter when wound with the magnetic wire, then the sheet used to make your first shroud should have a length of 9¾ inches × π = 30.6 inches plus another 2 inches for the overlap for the nylon bolts, or 32.6 inches long, with 15 inches being a good height for the inner shroud. The second, outer shroud then can be an inch wider in diameter and a couple inches shorter in height. Bend the strip of plastic into a cylinder held together with nylon nuts and screws spaced about 1 inch apart (Fig. 9-69).

Then place the Breakdown Baffle between the primary and secondary coils as in Fig. 9-70. Note in this figure that the red wire going to the Adjusting Tap originates from capacitor C1 (the attachment to the capacitor is not clearly seen but is on the right side of the large white capacitor C1, better seen in Fig. 9-64). The Adjusting Tap

Figure 9-69   Breakdown baffle.

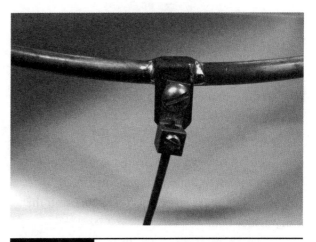

Figure 9-71   Strike ring and ground wire.

is then positioned on the Primary Coil to find the best resonant tuning. Also note the green grounding wire attached to the strike ring (see Fig. 9-71) and to the ground connection of the Secondary Coil (see Fig. 9-64).

The ground wire connection to the strike ring is made by bending a copper strip around the strike ring and soldering it in place and then securing an aluminum electrical block with a brass bolt and nut (Fig. 9-71).

Secure the No. 8 wire from the spark gap to the primary coil. Use whatever method works best for you—either a mechanical or soldered connection or both. Figure 9-72 shows a terminal block bolted onto a piece of copper flashing, which itself has been soldered onto the primary coil.

Figure 9-70   Completed base section of Tesla coil.

Figure 9-72   HV wire attached to start of primary coil.

Final Assembly: Putting It All Together

Figure 9-73 is a representative schematic showing how all the components and subassemblies are wired together.

Figure 9-74 provides an overview electrical schematic of how all the wiring comes together between the three main components: the contactor box and variac, the Tesla coil, and the transformer bank.

And finally, Fig. 9-75 shows a picture of how the Tesla coil would be wired up in reality, with the transformer array and power-control box at a safe operating distance. Note that the Tesla coil would not actually be *fired* in this particular setup because the drop cloth would be ignited and the door behind it likely would be damaged! When firing the coil, things should be at a safe distance (at least 15 ft) in all directions around the coil.

Testing Steps

The Fig. 9-4 and 9-74 schematics contain a 220-Vac 30-amp variac to control primary voltage.

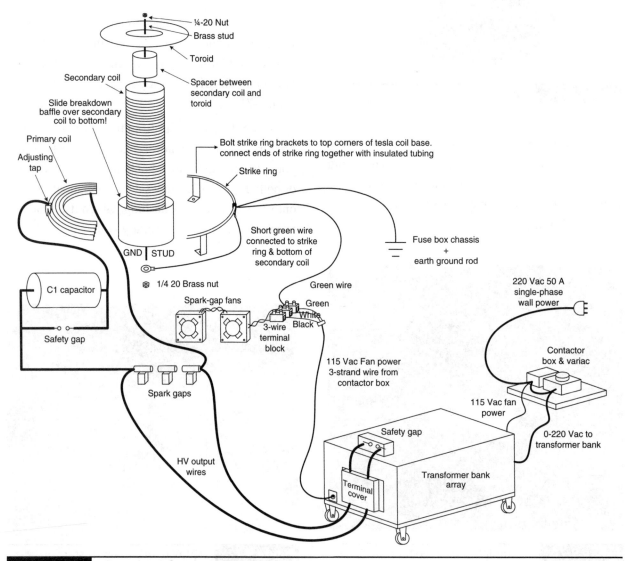

Figure 9-73   Overview of main component setup.

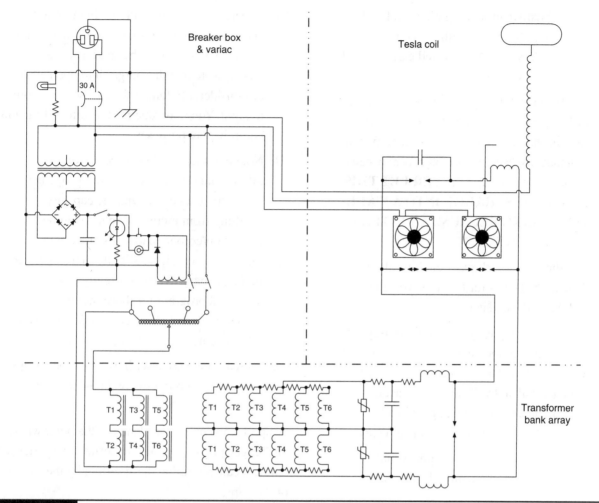

Figure 9-74   Overview schematic.

Figure 9-75   Power control in foreground with Tesla coil at a safe distance in background.

This is not required once the coil is tuned properly, but you should strongly consider it as part of the circuit because serious faults and mistuning, coupling, and so on can cause catastrophic results at full output during the setup and tuning phase. A variac also allows the coil to be operated at different outputs whenever desired.

1. If you live in a congested area, close to an airport, or near computers and other sensitive electronic equipment, it will be necessary to operate this system within a Faraday cage (see Chap. 20).

2. It is strongly advised that you test operate this system in an area with wooden or well-insulated floors to reduce accidental contact with dangerous ground currents.

3. Verify positive grounding of control and power supplies along with top connection of secondary coil. A dedicated earth ground should be used for a system this large.

4. Verify correct wiring and proper clearance of HV points. Verify position of coil, noting that discharges are possible up to 12 ft away and that strikes to certain objects may cause damage or start a fire. **NEVER LET THIS DEVICE DISCHARGE INTO A WALL OR CEILING BECAUSE AN INTERNAL FIRE CAN RESULT!**

5. Set the four spark gaps each to about 1/32 inch. This is for initial power-up and will be widened later.

6. Connect the tap lead to the outermost primary turn. You will note that the flexible primary tap lead will act as a partial turn of the LP1 primary coil and will add or subtract inductance depending on whether it is going with or against the primary coil winding direction.

7. Place a grounded contact approximately 12 inches from the terminal, and secure it in place.

NOTE Through extensive testing, we have determined the optimal vertical positioning between the primary and secondary coils for this design, so if you have followed all the design specifications, then the height of the primary coil has been set properly by the holding brackets.

8. Turn the unit on, and slowly rotate the variac just until the spark gaps fire. Note a discharge occurring to grounded contact. If it is not occurring, quickly turn the unit off, and recheck the system. Note that there is no interwinding breakdown because this indicates gross mistuning.

9. Separate the grounded contact to the point where spark is erratic, and connect the tap to the next inner or outer winding on the primary coil LP1. Reapply the power.

10. Repeat steps 8 and 9, attempting to find the exact point on the primary tap for maximum output. Slightly open the spark gaps, and repeat. Note that discharges into open air may be considerably longer than those from point to point. You may tune for this effect if the unit is for display purposes.

11. Note that there will be a second-order differential effect when the gap is opened owing to changes in dynamic capacity resulting from increased space-volume ionization/corona load. This effect will manifest as requiring slightly more inductance in the primary. We are currently developing an advanced program for those into the complex higher math and electrical physics of resonant HV systems.

12. Increase the variac settings, and repeat steps 8 to 12, attempting to obtain the longest discharge possible.

Spark gaps may be widened to the point where discharges just start to become erratic. Experienced builders can determine these setting by the sound of the gaps.

Our laboratory prototype easily produced 10-ft discharges into open air and grounded objects.

It is a good idea to keep notes as you experiment.

Special Notes

Note that you could use a radiofrequency (RF) ammeter to measure the secondary coil ground return current. This meter can be used for an indication of fine-tuning by adjusting to a maximum value. Initial operation where tuning may be "way off" can result in erroneous current peaks, especially if coupling is too tight. Use a thermocouple device if available.

The secondary coil in the system acts similar to a quarter-wave antenna with top-loading capacity. This means that there is a current node at the base in amps. If you depend on the green wire of the line cord for grounding, you force your electrical wiring to become a part of this system. This is not a good idea because voltage gradients and standing waves can result along the wiring run. This voltage value is determined by many complex factors such as frequency, harmonic content, and of course, varying amount of parallel and series complex impedances along "this run." What this means in simple language is: *Ground the coil to earth with the shortest dedicated direct lead possible!* This also can eliminate a good part of interference coupled back to the power line!

The coil form used for the secondary should be an excellent insulator and be a relatively lossless dielectric at the operating frequency. A ruby-mica coil form would be ideal if it existed. Material with this property is expensive. Polycarbonate is also good but expensive. Thin-wall PVC tubing, while not the best, is a good compromise between cost and performance. Unfortunately, PVC is hygroscopic (absorbs moisture). However, it can be treated by driving out the moisture with a heat gun or electric heater and then sealing it with orange shellac or equivalent (do this on a dry day).

CAUTION PVC tubing may give off undesirable toxic gases under HV stress. Lexan polycarbonate tubing is preferred if you are planning to use your coil a lot.

Not all holes on fabrication points are dimensioned. It is suggested on these to fit and fabricate as you go.

Safety spark gaps are connected across the transformer (see Fig. 9-30) and capacitor (see Figs. 9-43, 9-49, and 9-58) to limit overvoltage stress.

The spark gaps are fabricated from six pieces of 3- × 3/8-inch tungsten rod (or tool steel, but tungsten stays cleaner). A muffin fan keeps the assembly cool.

CAUTION Ultraviolet (UV) hazard. Spark-gap discharges emit hazardous ultraviolet emission and must not be viewed directly. Shield with plastic, or if you must observe, be sure to use a clear piece of plastic that blocks UV light or wear protective eyewear.

Tungsten is recommended for long-term reliable coil operation. However, tungsten is expensive, and units requiring short-term operation may use drill rod/tool steel. It will be necessary to reface tool-steel electrodes more often than if they were tungsten.

Special thanks to D. C. Cox, now deceased, at Resonance Research for his helpful suggestions and specially designed suppression circuit allowing use of neon-sign transformers for large coils.

Faraday Cage Assembly

Your Tesla coil is a resonant high-frequency transformer as well as a highly inefficient radio transmitter. As far as we know, after checking with the Federal Communication Commission (FCC), there has never been an interference report due to the use of these devices.

But shielding of your Tesla coil may be necessary depending on the strength of its RF output.

Tesla coils, even though with wound secondaries that approximate quarter wavelengths of the operating frequencies, are very poor RF radiators. However, some coils can produce radiated emissions in excess of 50 μV/m beyond the legal distance of 3 m. The problem usually can be solved with a reasonably easy-to-construct Faraday cage, as shown in Chap. 20.

Another issue is conducted emissions that can damage sensitive electronic equipment by entrance through shared 115-Vac wiring. This usually can be solved by installing an appropriate line filter directly to the Faraday cage. Dedicated grounding now must be done properly to resolve this problem.

Special Note on Selection of Primary Capacitor

There seems to be two schools of thought on the selection of this critical part.

Durlin Cox of Resonance Research suggested to select this capacitor on the high side of what is referred to as the resonance point, when combined with a neon transformer. This provides smoother running of the spark gap, with less current demand on the transformer, which is also now tuned above the resonant frequency. The downside is that as more capacitance is added to the system, the resonance condition is approached. This added capacitance can come from using a larger toroid placement of the coil such that it builds up capacitance, and more discharge output creating more Corona load capacity. Also, general experimentation could lead to reaching this point.

The other school of thought is to calculate the amount of capacitance that would equate out to the Resonant Rise condition and multiply it by 1.618. This now eliminates the problem of adding more capacitance. You would just force it further away from resonance. But on the other hand it does allow for more utilization of the transformer plus more output from the coil, at the cost of more heat and wear on the transformer.

We use the Durlin Cox method, and have never had a failure using over 500 NST in our coils.

Ref. No.	Quantity	Description	Part #
TABLE 9-1 Parts List for Power Supply*			
			Power Supply
R1		39-KΩ ¼-W film resistor (ONG, WHI, ONG)	
R2		1-KΩ ¼-W film resistor (BRN, BLK, RED)	
C1		1000-µF 50-V vertical electrolytic	
C2, 3		1-µF 250-Vac metalized polypropylene	
D1		1N4001 1-amp 50-v rectifier diode	
NE1		Neon indicator lamp	
LED1		Green led	
T1		220-Vac in, 24-vac 1 amp out	TR220/24/1A
VA1		220-Vac 30-amp in, 0–250-vac 8-kvac out	VARIAC220V30A
BK1		30-amp two-line breaker	BK220V30A
RL1		P31-E1014-1 coil 24-vdc 35-amp, 600-vac 50/60-hz	RELAY600V35A
BR1		RB151 50-V 1.5-amp bridge rectifier	RB151
S1		1-amp key switch S106-1	KEYSWSM
S2		1-amp push button	
P1		220-V 50-A welder plug	
J1		1/8-inch stereo jack	
			Enclosures
Top		22- × 9- × 1/16-inch aluminum	
Front		10- × 8- × 1/16-inch aluminum	
Rear	2	3¼- × 7¾- × 1/16-inch aluminum	
T cover		6¼- × 8½- × 1/16-inch aluminum	
		Contactor box with 30-amp two-line breaker capability	
		Contactor bracket	
		Power supply main base	
		Assembled Contactor Box (Fig. 9-5) available as a ready-to-use module	BTC7-CB
		Fully assembled Power Supply (Contactor Box, Variac, Remote Switch; Fig. 9-9) available as a ready-to-use module	BTC7-PSUV

*Most parts should be available through electronics or hardware stores, but those more difficult to acquire are listed with a "Part #" and are available through www.amazing1.com if needed.

TABLE 9-2	Parts List for Transformer Box*		
Ref. No.	**Quantity**	**Description**	**PART #**
		Transformer Box	
		Box Base	
		2-ft × 3-ft × ¾-inch plywood, two layers	
	4	2-inch caster wheels	
Side rails		35-inch 2 × 4 ripped in half for two rails	
In/out brkt	2	2½- × 4- × 1/16-inch galvanized steel	
		Box Lid	
Top		2-ft × 3-ft × ½-inch plywood	
Wall	2	11½-inch × 3-ft × ½-inch plywood	
Faces	2	11½- × 23- × ½-inch plywood	
Brace		7-ft 2 × 4 ripped in half for internal bracing	
Vent cover		1-inch vent hole cover	
		Transformer Circuit	
T1-6	6	15-kV 60-milliamp current-limited non-GFI transformer	15KV60MA
R1-10	10	300-Ω 50-W wire-wound resistor	300/50W
Ground		1- × 27½- × 1/16-inch aluminum	
		3 ft no. 12 wire from transformers to suppression circuit	
		3 ft 20-kV HV wire from transformers to suppression circuit	WIRE20KV
		Suppression Circuit	
SR1–4	4	300-Ω 50-W wire-wound resistor	300/50W
SC1, 2	12	6× 0.033-µF 1.6-kV	.033u/1.6KV
SMOV1, 2	16	8× 1800-V metal oxide varistor	MOV1800
SL1, 2	2	80-T no. 24 magnet wire wound on 2-inch OD PVC tube	COIL80T-24SUP
Layout BRD		9¼- × 10- × 1/8-inch polycarbonate	
Ins. sheet		10- × 11- × 1/8-inch polycarbonate	
		¾-inch spacer sleeves	
		Solder lugs	
		4 ft GTO HV wire from suppression circuit to HV bracket to safety gap	WIRE40KV
		Assembled Suppression Circuit (Fig. 9-15) available as a ready-to-use module	BTC7-SUP
		Safety Gap	
Base		2- × 7½- × 1/8-inch lexan sheet	
Bracket		½- × 2- ×1/16-inch copper or aluminum bent at 3/4 inch	
	2	½-inch-diameter brass ball 6-32 internal thread	BALL12
		6-inch 6-32 threaded rod	
		Solder lugs	
	2	Wire nut	
		HV Bracket	
Bracket leg		3- × 8- × 1/16-inch aluminum cut diagonally for two legs	
Bracket		8- × 11- × 1/8-inch lexan	
	2	Terminal lugs	
	2	Terminal blocks	
Cover		8½- × 6¼- × 1/16-inch plastic	
		Fully assembled Transformer Box including suppression circuit, safety gap, HV output (Fig. 9-22, 9-30), but not including power supply, available as a ready-to-use module	BTC7-TXBX

*Most parts should be available through electronics or hardware stores, but those more difficult to acquire are listed with a "Part #" and are available through www.amazing1.com if needed.

TABLE 9-3 Parts List for Coil Assembly*

Ref. No.	Quantity	Description	Part #
		Coil Assembly	
		Chassis	
Base		2-ft × 2-ft × ¾-inch plywood	
Feet	4	2-inch PVC tube cap	
Leg	4	15- × 1-5/8-inch OD clear plastic or PVC	
Leg cap	8	1-5/8-inch plastic cap	CAP158
Leg center	4	17 inches 1/4-20 threaded rod	
Top plate		2-ft × 2-ft × ¾-inch plexiglas or plywood	
		Three- wire terminal block for fans and ground	
		Capacitor Assembly	
Side		4- × 8- × ¾-inch plywood	
Wall		2- × 12- × ¾-inch plywood	
C1		0.05-µF 20-kVac special oil-filled polypropylene capacitor designed for tesla coils	.05u/20KVac
		Safety Gap	
Base		2- × 7½- × 1/8-inch lexan sheet	
Bracket		1/2- × 2- × 1/16-inch copper or aluminum bent at ¾ inch	
	2	½-inch-diameter brass ball 6-32 internal thread	BALL12
		6 inches 6-32 threaded rod	
		Solder lugs	
		Wire nut	
		Spark Gap	
Base	2	3½- × 6¼- × 1/16-inch lexan	
	6	1¼- × 1- × ¾-inch pvc block	BLK125PVC
	6	1¼- × 2¼- × 1/16-inch galvanized steel	
	8	3/8-inch terminal block lug	LUGBLK38
	6	3/8- × 2-inch tungsten	TUNG3825
	4	Terminal lug	
Fan	2	4.69-inch² × 1½-inch axial fan	FAN115-4
Fan shroud		16- × 5-7/8- × 1/16-inch black plastic	
		Assembled spark gap (Fig. 9-54), with cover, available as a ready-to-use module	BTC7-SPKGAP
		Primary Coil	
PRI BRKT	8	4- × 7- × 1/16-inch lexan	
PRI coil		Roughly 60-ft × ½-inch × 0.02-inch copper ribbon	
PRI tap A		4- × ½- × 0.02-inch copper ribbon	
PRI tap B		3- × 1- × ½-inch wooden or PVC block	
PRI leads		7 ft no. 8 stranded wire	WIREFLEX8

*Most parts should be available through electronics or hardware stores, but those more difficult to acquire are listed with a "Part #" and are available through www.amazing1.com if needed.

Ref. No.	Quantity	Description	Part #
TABLE 9-3		Parts List for Coil Assembly (*Continued*)	
			Coil Assembly
			Breakdown Baffle
		33- × 15- × 1/16-inch plastic	
		34- × 13- × 1/16-inch plastic	
		Ground strike ring	
		60- × ¼-inch OD copper tube	
		6- × ¼-inch ID plastic connection TUBE	
		3/8-inch terminal block	
GS BRKT	4	10- × 2- × 1/16-inch lexan bent at 2 inches	
		4 ft no. 12 green stranded wire	
			Secondary Coil
		4-ft × 8-5/8-inch white PVC	
		2400 ft no. 18 magnet wire	
	2	8- × 3- × ½-inch PVC cut and shaped to fit	
Spacer		4½- × 6-inch OD clear plastic or PVC	
		6-inch 1/4-20 threaded brass rod	
		30-inch seamless toroid terminal	
		Assembled Secondary Coil LS1 (Fig. 9-38) available as a ready-to-use module	BTC7COIL

CHAPTER 10

Solid-State Tesla Coil

Overview

This solid-state Tesla coil (Fig. 10-1) is used as a spectacular display of lightning-like discharges for museums, attention-getting advertising, special events, and of course, high-voltage, high-frequency research. With the audio interface circuitry, a keyboard can be attached such that the electrical arc frequencies resonate as the notes of the keys being played, and you can actually play music right through these electrical arcs!

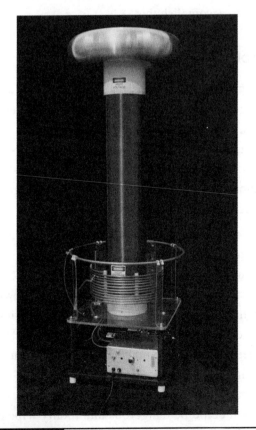

Figure 10-1　Solid-state Tesla coil.

Solid-state Tesla coils have been around for nearly 40 years. In the past, they used vacuum tubes providing oscillations up into the 100- to 500-kHz range. These frequencies allowed the use of solenoid coils within practical sizes. The vacuum-tube coil is very forgiving to design errors, electrical breakdowns, and other faults because tubes take much more abuse than a semiconductor. These coils, when designed properly, produced impressive results but use extremely dangerous plate voltage transformers that could electrocute or seriously harm an inexperienced experimenter.

True solid-state circuitry for Tesla coils was first achieved by Ritchie Burnett, a U.K. electronics engineer. His designs were continuously operating, and while the display was energetic, the spark lacked the length of more current designers using duty-cycle synchronized pulse control. Now the energy could be compressed into time intervals that allow the solid-state Tesla (SST) switches to supply higher peak power yet use less Joulean energy. This approach takes advantage of the peak-current capability of the insulated-gate bipolar transistor (IGBT) switches.

SST circuit designs have evolved into a single resonant mode where the secondary coil is the resonant control of the switching frequency. Coupling is now a non-resonant-matching section. This method works well, allowing continuous operation, providing impressive displays, and improving modulation so that the arc can produce clearer sounds.

The QUASAR60 double-resonant design is where the primary is also resonant with the secondary, and power transfer requires less coupling. Because of the high *Q* of the system, immense power is easily transferred and would destroy switches and associated circuitry if allowed to operate only for a millisecond. Duty-cycle switching controlling the switching time to a fraction of the frequency period must be used.

We took some interesting video on the arc discharge of the coil that occurred for 2.3 seconds in the continuous mode. This discharge was a yellowish orange with intense heat that was felt several meters away. It could best be described as a sheet of a pure plasma flame. A hole was burned into the toroid after the needle was vaporized. It was weird and unlike anything we have seen. We have 440-V three-phase power at our lab, and the breakers tripped during this event, which did not make sense because they are well over 100 amps. Note that we used a feedback (FB) driver using some superpowered field-effect transistors (FETs) with 1500 V on the drains—they survived, thank the Lord, because they cost over $2500 each. This driver is for an earthquake-simulating device.

Steve Ward, Dan McCauly, and many other experimenters were mainly responsible for many solid-state Tesla coil designs. Dan McCauley's book on solid-state Tesla design is a collection of many of these designs interfaced into a professionally engineered, reliable, and highly impressive vehicle. We strongly suggest that any Tesla Coil hobbyists obtain Dan's book because it is a wealth of information and by far the best we have seen on this subject.

Our QUASAR60 coil uses Dan McCauley's driver with some modifications. His book was a great aid to our designer, John Wilford, who created our final prototype.

Hazards

Even though this Tesla coil has a high-voltage output, unlike the transformers used in standard core-and-coil designs, this advanced device poses a less dangerous or lethal electrical hazard—which

is one of the attractive features of solid-state Tesla coils. But still, anything that plugs into 115-Vac wall power poses an electrical hazard, and precautions for working with high-voltage items should be observed. Eye protection should be worn when making, testing, and operating this device. Ear protection may be preferred, but this solid-state coil is not as loud as the core-and-coil BTC70 from Chap. 9—and if this is run with an audio input, such as a keyboard, you will want to hear the notes produced by the modulated arc.

Difficulty

Requires *advanced* skills in wiring and soldering, along with *experience* in fabricating sheet metals and plastics and woodworking. Electronic knowledge in the fields of high-voltage (HV) and radiofrequency (RF) devices should be familiar to the builder. Construction is shown as individual modules, each being built and tested. A coil must be wound and labor will benefit from a special lathe to accomplish this.

Tools

Wiring tools of several sizes, both a 50-W and a larger 100-W soldering iron, hand and power tools, fabrication equipment, good voltmeter and ammeter, frequency generator, and a *low-cost oscilloscope*. Electronic ability to use test equipment such as meters and scope will help in understanding and completing this project.

You should familiarize yourself with Chap. 21 before you build this circuit or any similar circuit without an isolation transformer. We advise you to use the testing-circuit jig shown at Chap. 21 for this project because it verifies dangerous alternating-current (ac) grounds. Other amenities are powering a virgin circuit starting with low input voltage and slowly increasing the voltage, noting that any excessive current could be dangerous and totally wipe out your hard work!

Secondary Coil Notes

Extensive calculations and experiments have shown us that the best results in a secondary coil are 1400 turns of 24 gauge wire around a 6.5-inch diameter tube of 40 inches overall length, with 3 inches offset on the coil bottom, then 29 inches of vertical winding distance, with a remaining 8 inches on top of the secondary coil (see section 5d). You can experiment with different setups, but we have found this to be optimal.

Assembly

Compared with the traditional core-and-coil Tesla coil (such as the BTC70 of Chap. 9), the solid-state Tesla coil has more circuitry in the design but less hardware in the power supply.

The assembly of this solid-state Tesla coil can be broken down into six parts:

1. Full bridge and tank circuit
2. Controller
3. Interrupter
4. Power supply rectifier and variac
5. Chassis (base, strike ring, primary coil)
6. Secondary coil + output terminal

1. Full Bridge and Tank Circuit Assembly

Figure 10-2 shows the full bridge with tank circuit schematic.

R1-6 10-MΩ 1-W carbon or silicon

R7-14 10-Ω 1-W film resistor

C1–12 0.15-μF 2-kVdc 600-Vac

C13,14 1-μF 600-V

C15 6300-μF 400-Vdc

D1–4 IN5819

D5–12 IN4749

D13–16 1.5KE30CA

D17–20 MUR1660

D21–24 1.5KE400CA

Q1–4 HGT1N60N60

Note that the 12 capacitors shown in the Tank Circuit do not have much wiggle room for voltage standoff. Our testing coil has used them for 6 months without failure, even though the series resonant voltage rise has gone well above their ac rating. A safer approach would be to use a 0.33-μF capacitor at 600 Vac, now providing a voltage standoff of 7200 Vac at the same capacity—but requiring twice the space.

1a. Full Bridge and Tank Circuit Assembly: Full Bridge

Figure 10-2 shows the full bridge with tank circuit schematic. First, cut the foundation sheet from ¼-inch-thick Lexan or plastic with dimensions of approximately 3¾ × 13½ inches (Fig. 10-3). Drill holes at the corners that will be used later to secure this sheet to the wooden base plate—these don't have to be precise, but setting them back about ¼ inch from the edges should be fine.

Next, fabricate the aluminum heat sink as shown in Fig. 10-4. Drill and tap the two holes at the bottom leading edge of the heat sink for 6-32 threads (use 1¼-inch 6-32 screws to hold the fan) placed at whatever location will best hold the cooling fan. The eight holes on the top of the heat sink are 1/8 inch in diameter, 5/16 inch deep, and untapped because the No. 6 "blunt" self-tapping screws will form their own threads in the soft aluminum.

Next, screw the IGBTs to the heat sink (Fig. 10-5) using No. 6 "blunt" screws of ¼ inch length. If the holes were drilled properly to 1/8 inch diameter, the screws should go in with minimal resistance. But be careful if there is too much resistance (generally, if the screws become difficult to turn) because the heads may snap off. Thread all eight holes first before attaching the IGBTs. This is easier and prevents the screwdriver from accidentally slipping

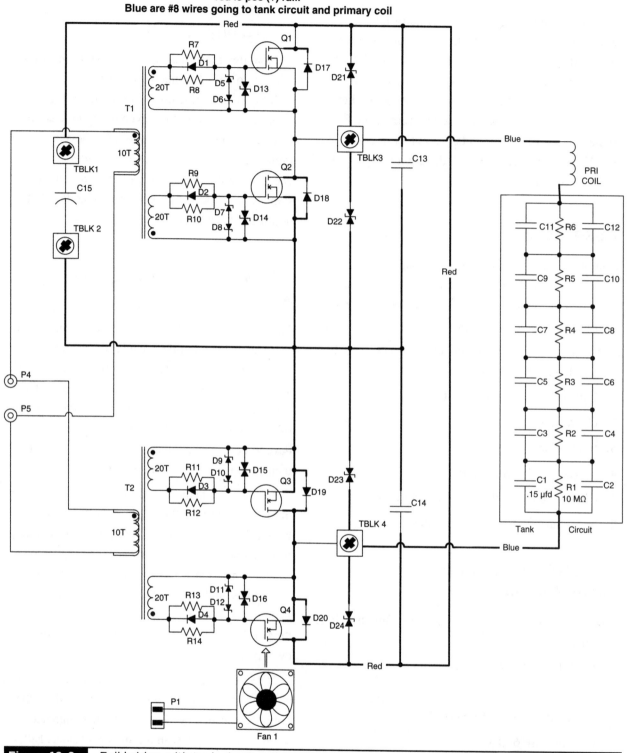

Figure 10-2 Full bridge with tank circuit sche matic.

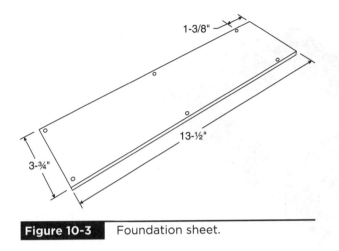

Figure 10-3 Foundation sheet.

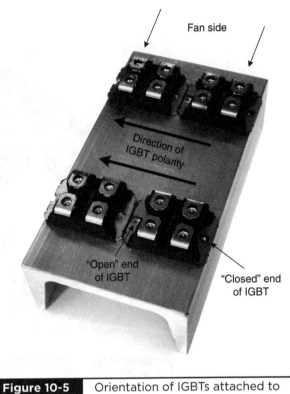

Fan side

Direction of IGBT polarity

"Open" end of IGBT

"Closed" end of IGBT

Figure 10-5 Orientation of IGBTs attached to heat sink.

and damaging the IGBTs. Once the eight holes are threaded, attach the IGBTs to the heat sink with the orientation shown in Fig. 10-5 because they have a polarity (the screws are left out of this figure to better show the IGBT orientation, but screw them in place once they are oriented properly). Remember that the screws are threading into aluminum, so do not

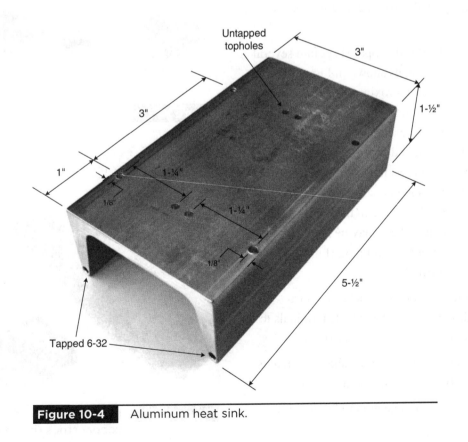

Untapped topholes

3"

1-½"

3"

1"

1-¼"

1/8"

1-¼"

1/8"

5-½"

Tapped 6-32

Figure 10-4 Aluminum heat sink.

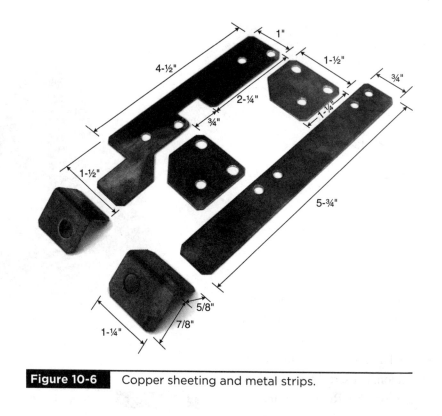

Figure 10-6 Copper sheeting and metal strips.

overtighten them into this soft metal! Just slightly snug is fine.

Now cut the 20–30-mil copper sheet into several pieces with the dimensions shown in Fig. 10-6. The hole locations are approximate—the most important thing is that they line up with the holes of the IGBTs, so drill these holes to match the actual physical layout, and test-mount the strips on the IGBTs with the screws in place to verify proper alignment. The two bent copper strips that attach to the capacitor should line up with the copper sheets on the IGBT/heat sink; these may be notched to allow a range of vertical positioning (Fig. 10-6 shows them only drilled).

Clip the center ground terminal of the fast-recovery diodes, and bend the outside terminals to lay flush with the copper strips (Fig. 10-7).

Attach the diodes to the copper sheets with nuts and bolts (Fig. 10-8 shows the bottom view).

Bend 20 electrical connectors to 90 degrees (Fig. 10-9).

Figure 10-7 Clip and bend the diodes.

Figure 10-8 Fast-recovery diodes attached to copper strips (bottom view).

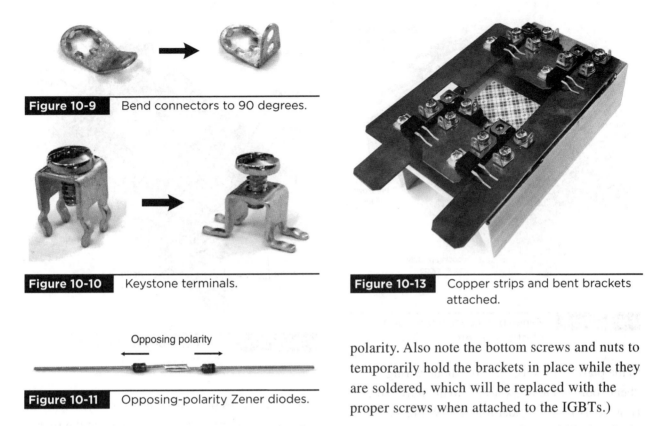

Figure 10-9   Bend connectors to 90 degrees.

Figure 10-10   Keystone terminals.

Figure 10-13   Copper strips and bent brackets attached.

Opposing polarity

Figure 10-11   Opposing-polarity Zener diodes.

Bend out the legs of four screw-type keystone terminals, and open the top screw (Fig. 10-10)

Solder together four pairs of Zener diodes with opposing polarity (Fig. 10-11).

Cut 2 strips of 1/16-inch plastic sheeting measuring ½ × 4½ inches each, bend and drill, and mount 4 electrical brackets to each by screws and friction nuts and then solder 4 resistors R7-14 and 2 diodes D1-4 to each (Fig. 10-12). (Note electrical bracket orientation and diode

polarity. Also note the bottom screws and nuts to temporarily hold the brackets in place while they are soldered, which will be replaced with the proper screws when attached to the IGBTs.)

Now attach the 4 copper strips and 12 electrical brackets to the IGBTs; also put down some double-sided sticky foam to hold one of the large white capacitors (Fig. 10-13). Note the orientation of the electrical brackets.

Next, some soldering. First, screw the two plastic brackets (with their own attached electrical components) into place across the IGBTs. Then solder four 30-W breakover diodes across the electrical brackets parallel to the four pairs of Zener diodes (of Fig. 10-11) beneath and adjacent

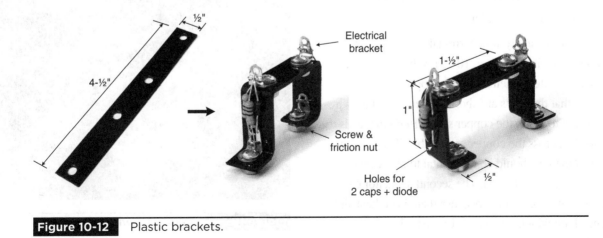

½"

Electrical bracket

4-½"

1-½"

1"

Screw & friction nut

Holes for 2 caps + diode

½"

Figure 10-12   Plastic brackets.

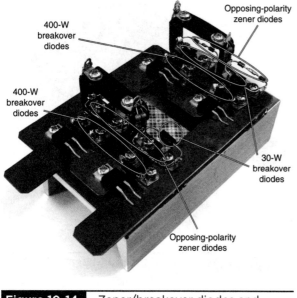

400-W breakover diodes

Opposing-polarity zener diodes

400-W breakover diodes

30-W breakover diodes

Opposing-polarity zener diodes

Figure 10-14   Zener/breakover diodes and plastic brackets attached.

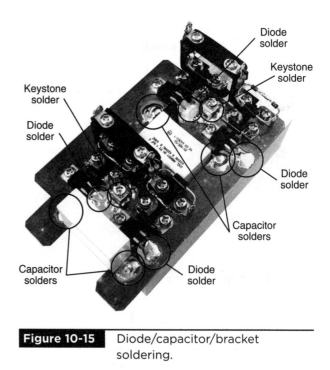

Diode solder

Keystone solder

Keystone solder

Diode solder

Diode solder

Capacitor solders

Capacitor solders

Diode solder

Figure 10-15   Diode/capacitor/bracket soldering.

to the plastic brackets. Note the diode wiring "bent" outward to allow access to the IGBT screws (Fig. 10-14). Then solder four 400-W breakover diodes across the remaining electrical brackets.

When soldering the two-terminal diodes D17-20 to the copper strips, it is best to heat the copper strip with a wide-head 100-W gun to transfer heat quickly. Touch the full tip across the copper, apply a little solder between the tip and the copper to spread out and give a fast heat transfer, and after several seconds the copper should be hot enough to quickly add more solder, make a small puddle, and then remove the heat tip. Do not heat the diode terminals directly! If the terminals are heated, the diodes will melt before the copper becomes hot enough to melt the solder.

Also solder a keystone terminal on each center copper strip "island" (Fig. 10-15). When soldering the large white capacitors to the copper strip, it may happen that the leads are short and require a bus-wire extension to reach the copper strips. Bend the leads to lie as flat across the copper as possible. The capacitor nearest the fan should be horizontal to allow better air flow (see Fig. 10-15). The second capacitor can be stood up "vertically" because there is no heat sink beneath on which it can rest (Fig. 10-15).

Cut 48 inches of four-strand 22-gauge telephone wire, and strip the outer jacket such that the four inner wires (black, yellow, red, and green) are exposed. *Tip:* Cutting the entire jacket is tedious, so a nice method is to just cut open about 1 inch of the jacket, then hold the four wires in a vice, and pull the jacket off (Fig. 10-16). Be sure to pull from the top down, which will create a kind of "bubble" in the jacket as you go. (If you pull

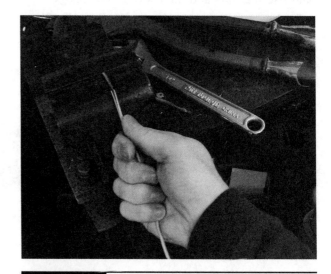

Figure 10-16   Removing wire jacket.

from the bottom, the jacket will just tighten up like a Chinese finger trap, so the trick is to create a looseness in the jacket to slide it off.) The jacket will not be pulled off in one swoop, but rather your fingers will need to slide down the length of the jacket several times, pulling the bubble somewhat tightly as you go. With a little practice, a jacket can be removed in this way in a few seconds, but you only need to strip two.

Once the outer jacket is removed, start twisting the four strands of wire together (it needn't be overly twisted, maybe about 1 turn per inch or so), then cut the black and yellow wires after about 28 inches of twisted 4-strand wire, and keep twisting the remaining 20 inches of 2-strand red/green wire until done. Make one more of these.

Leave about 6 inches of the four-stranded wire hanging free out of the core; then wind the four-strand wire 10 times through the core, keeping the wire snug and close to the core as you go. After the tenth winding, separate the remaining length of black and yellow wires (there should be about 6 inches of black and yellow wire on the opposite side) from the four strands (Fig. 10-17).

Space the four-strand windings evenly around the core to give space between each winding (Fig. 10-18).

Then continue winding the red and green wires another 10 times around the core, going between the four-stranded windings (Fig. 10-19). This completed assembly is now a Gate Transformer. Make one more of these.

Figure 10-18   Equidistant spacing of four-strand winding.

Use tie wraps to secure the Gate Transformer coils to the top of the plastic brackets, with the 4-strand wire side of each coil facing the capacitor and the 2-strand wire side of each coil facing the fan (Fig. 10-20).

Solder the red and green wires to the electrical connectors as shown in Fig. 10-21. Looking at the fan side of the heat sink, the green wire coming *toward* the fan attaches to the upper-left connector on the plastic bracket, and the red wire coming *toward* the fan attaches to the bottom-right IGBT on its' left connector. The green wire going *away* from the fan attaches to the bottom-left IGBT on its' left connector, and the red wire going *away* from the fan attaches to the upper-right connector on the plastic bracket. (See www.amazing1.com/eg3 if more clarity is needed.)

Epoxy the aluminum heat sink to the plastic foundation. Leave about 1 inch from the end to allow space for the cooling fan (Fig. 10-22).

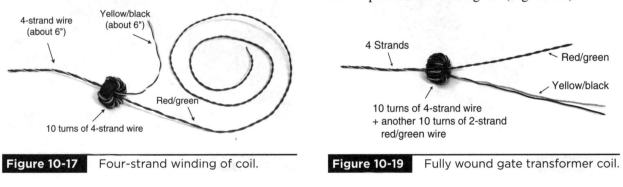

4-strand wire
(about 6")

Yellow/black
(about 6")

Red/green

10 turns of 4-strand wire

Figure 10-17   Four-strand winding of coil.

4 Strands

Red/green

Yellow/black

10 turns of 4-strand wire
+ another 10 turns of 2-strand
red/green wire

Figure 10-19   Fully wound gate transformer coil.

Figure 10-20 Gate transformer coils secured to plastic brackets.

Figure 10-22 Heat sink epoxied to foundation.

Next, attach the Gate Transformer coil wires to the Input and Return plugs.

Input plug. The two sets of wires from the "inner" side of the coils should be soldered to separate (inner and outer) posts on the

Input plug (Fig. 10-23; the left plug colored red): Two wires are soldered to the inner post and two to the outer post. It doesn't matter which wires attach to which post as long as one set of black and yellow wires attaches to one post and the other set of black and yellow wires attaches to the other post. This Primary Current plug later goes into jack J3, the Primary Current jack on the front of the Control Box (see Fig. 10-41).

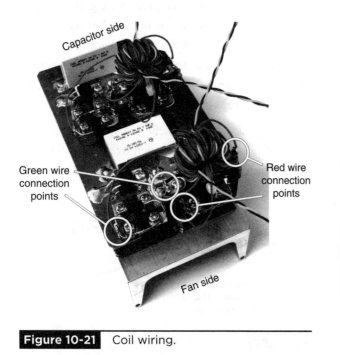

Figure 10-21 Coil wiring.

Figure 10-23 Gate transformer plugs.

Return plug. The two sets of wires from the "outer" side of the coils are soldered to the outer sleeve of the Feedback plug (see Fig. 10-23). This plug later goes into Feedback jack J4 on the back of the Control Box (see Fig. 10-42).

Once the heat sink has dried on the base, attach the large capacitor. First, make note of the capacitor's correct polarity alignment: With the heat-sink fan on the right side, the capacitor's negative terminal should be toward you (Fig. 10-24).

Attach the L-brackets to the capacitor, and then lay the capacitor on the foundation sheet (again, with the negative terminal facing you because the heat-sink fan is on the right). Adjust the L-brackets to be in full contact with the copper sheets with some small clamps (or even paper clips will work). Snug up the screws on the capacitor by hand, which may take some fiddling around because the working area is tight. Now that the L-brackets are aligned, remove the clamps holding them to the copper strips; then run a bead of epoxy along the bottom of the capacitor and another along the foundation, and put the capacitor back into place on the foundation sheet (use the clamps again on the L-brackets to maintain a good contact while the epoxy dries). Remember that after the epoxy dries, the L-brackets can still be adjusted, but it is best to have as good a start as possible.

Solder the L-brackets to the copper strips with a keystone terminal on each, and the Full Bridge now should be complete (Fig. 10-25).

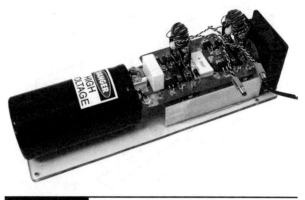

Figure 10-25 Assembled full bridge.

1b. Full Bridge and Tank Circuit Assembly: Tank Circuit

This is a relatively simple construction involving 12 capacitors in series parallel and 6 resistors to balance capacitor voltages (see the schematic in Fig. 10-2). Note that the 12 capacitors do not have much extra room for voltage standoff. However, our testing coil has used them for 6 months without failure, even though the series resonant voltage rise has gone well above their ac rating. A safer approach may be to use 0.33-µF capacitors at 600 Vac, providing a voltage standoff of 7200 Vac at the same capacity, but this would require twice the space (caution, as this has not been tested!).

Make a plastic holding chassis and cover, bent from 1/16-inch plastic sheet (Figs. 10-26 and 10-27).

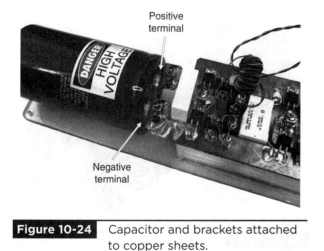

Figure 10-24 Capacitor and brackets attached to copper sheets.

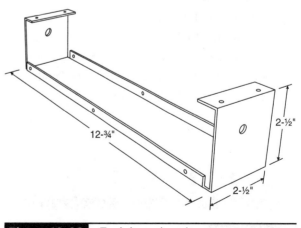

Figure 10-26 Tank box chassis.

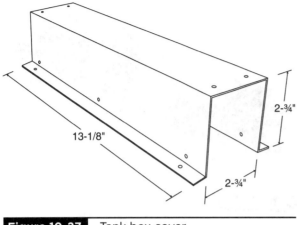

13-1/8"

2-¾"

2-¾"

Figure 10-27   Tank box cover.

Drill 1/8-inch holes to align the chassis and cover (the easiest way is to first bend the chassis and cover to shape and then assemble the two halves together and drill the holes in place such that everything is aligned properly). After the circuitry is fabricated, No. 6 "blunt" self-tapping screws will hold the cover in place. Also drill ¼-inch front and back holes in the chassis for the HV terminals (explained later).

Solder the components onto a 12¼- × 2¼-inch perf board (see Figs. 10-28 and 10-29 for general layout). Note that the resistors cannot be seen in Fig. 10-28, but they are there between the capacitors (as per the schematic in Fig. 10-2). Solder No. 12 wire to each end of the perf board, and then crimp these to an electrical terminal.

Attach strips of 1/8-inch-thick dual-adhesive foam to offset the solder joints and hold the capacitor array to the chassis (Fig. 10-29). Secure the HV terminals to the chassis with 1-inch ¼-20 brass screws and nuts (see Figs 10-28 and 10-30; but short straight wires are better than the "coiled" ones shown, to avoid inductance).

Figure 10-28   Tank box chassis and circuitry.

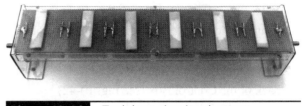

Figure 10-29   Tank box circuitry bottom.

Then attach the cover to the chassis for the completed Tank Box (Fig. 10-30). The Tank Box then is mounted to the base and wired between the primary coil start and primary coil tap (see Figs. 10-98, 10-102, and 10-103).

With the Full Bridge and Tank Box complete, the Controller is assembled next.

2. Controller Assembly

The circuitry (Fig. 10-31) for the Control Box is shown laid out on a standard perf board (Fig. 10-32), but a printed circuit board (PCB) is also available and can be found on our website at www.amazing1 .com (Item PCQUASAR60-CB) that will provide an easier build for those who want it.

Control Board Wiring Connections

OPTICAL: fiber-optic cable going to OPTO BUSH on Controller Box front panel (see Figs. 10-38 and 10-41). The fiber-optic

Figure 10-30   Tank box.

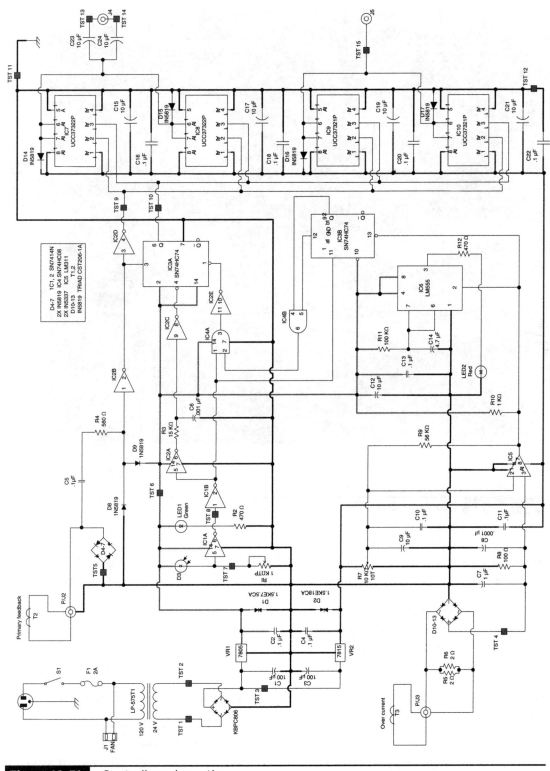

Figure 10-31 Controller schematic.

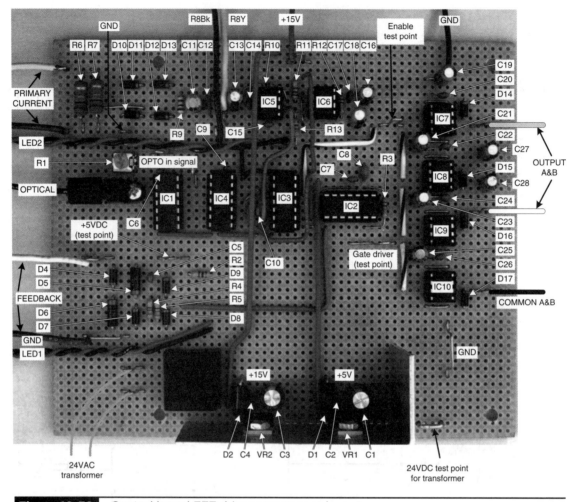

Figure 10-32 Control board FET driver component layout.

cable attaches here from the Interrupter (see Fig. 10-45).

LED1: red and black wire going to green LED1 (which indicates power ON) on Controller Box front panel (see Figs. 10-38 and 10-41).

LED2: red and black wire going to red LED2 (which indicates overcurrent) on Controller Box front panel (see Figs. 10-38 and 10-41).

PRIMARY CURRENT: yellow and green wires going to Primary Current jack J3 (yellow to center post, green to sleeve) on front panel (see Figs. 10-38 and 10-41).

FEEDBACK: white and green wires going to Feedback jack J2 (white to center post, green to sleeve) on front panel (see Figs. 10-38 and 10-41).

R7Bk: black wire going to Current Control dial R7 on front panel (see Figs. 10-38 and 10-41).

R7Y: yellow wire going to Current Control dial R7 on front panel (see Figs. 10-38 and 10-41).

+15V: red wire going to Current Control dial R7 on front panel (see Figs. 10-38 and 10-41).

GND: black wire going to chassis ground.

OUTPUT A&B: yellow and white wires going to jack J5 (yellow to sleeve, white to center post) on rear panel (see Figs. 10-39 and 10-42).

COMMON A&B: black wire going to jack J4 (sleeve) on rear panel (see Figs. 10-39 and 10-42).

24-VAC TRANSFORMER: blue wires going to transformer T1 (see Fig. 10-40).

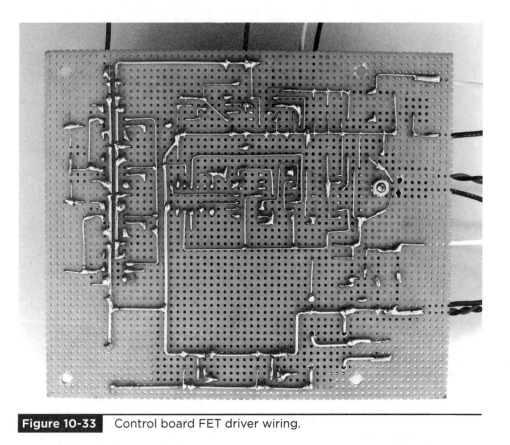

Figure 10-33   Control board FET driver wiring.

The Control Board is a somewhat complicated circuit, so two other views (Figs. 10-34 and 10-35) are included to help clarify any questions with the wiring that may arise Free full-color HQ images are also available at www.amazing1.com/eg3.

Use a piece of aluminum for the Control Board heat sink (Fig. 10-36).

Fabricate the Controller Box by bending and cutting a 1/16-inch aluminum sheet (or plastic may

Figure 10-34   Control board component layout and wiring.

Figure 10-35   Control board component layout and wiring.

be used for the cover) to the dimensions shown in Fig. 10-37. Alternately, a ready-made metal box (such as an electrical box) may be used with similar or larger dimensions.

Then cut an insulating sheet from 1/16-inch Lexan or similar plastic to fit into the base of the Controller Box. This sheet will insulate the exposed circuitry of the Control Board from grounding to the metal box. The insulating sheet is secured in place with the nuts and bolts going through the Control Board and transformer (see Fig. 10-40)—

the holes for these bolts do not need to be drilled into the Controller Box ahead of time but rather can be drilled as needed when the Control Board and transformer are mounted in the box.

Mount and wire the Controller Box front-panel components (Fig. 10-38) and rear-panel components (Fig. 10-39).

Controller Front-Panel Wiring Connections

(Refer to Fig. 10-32 for LED2, R7, J3, and LED1 wire connections.)

S1Bk: black wire going from switch S1 to rear outlet J1 (see Fig. 10-39).

PWR-W: white wire coming from PWR input wire and going to rear outlet J1 (see Fig. 10-39).

PWR-G: green wire coming from PWR input wire and going to chassis ground (see Fig. 10-39).

PWR-F1: black wire going between PWR input wire and fuse F1.

S1-F1: black wire going between switch S1 and fuse F1.

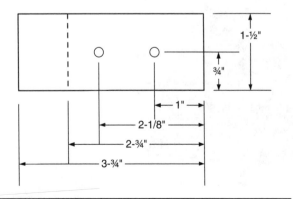

Figure 10-36   Control board heat sink.

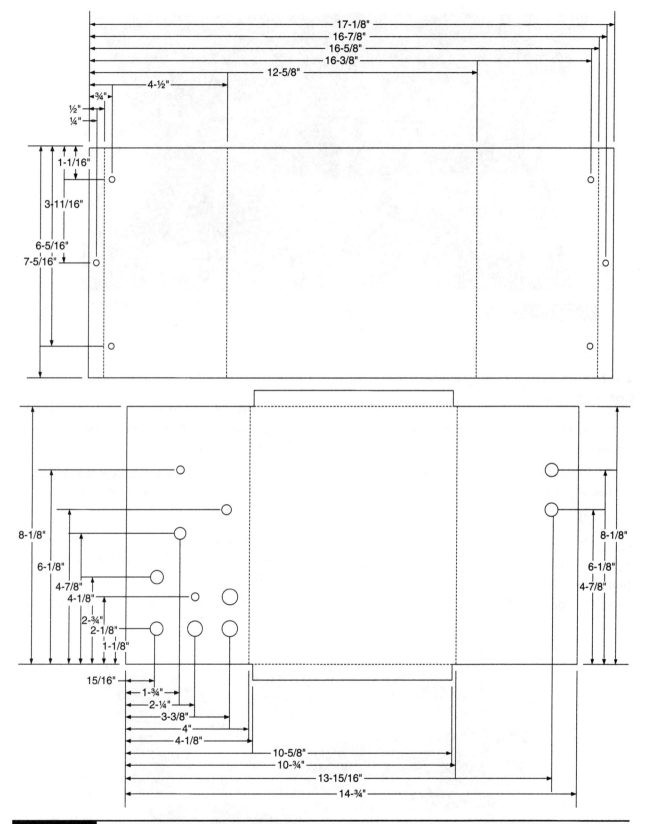

Figure 10-37 Controller box cover and chassis.

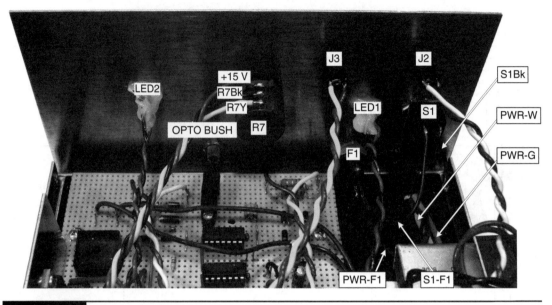

Figure 10-38 Controller front-panel components.

Controller Rear-Panel Wiring Connections

(Refer to Fig. 10-32 for J1, J5, and J4 wire connections.)

T1-R: two red wires going to transformer T1 (see Fig. 10-40).

T1-Bk: two black wires going to transformer T1 (see Fig. 10-40).

Figure 10-40 shows the chassis inside layout with mounted and wired components, and Fig. 10-41 shows the Controller Box front panel.

Controller Box Front-Panel Jacks

J2: feedback jack to accept the feedback plug from the Current Transformer (see Fig. 10-73).

J3: primary current jack to accept the primary current plug from the Current Transformer (see Fig. 10-73).

OPTO BUSH: accepts the Optic Cable AO-Bk from the Interrupter Box (see Figs. 10-45 and 10-48).

Figure 10-42 shows the Controller Box rear panel.

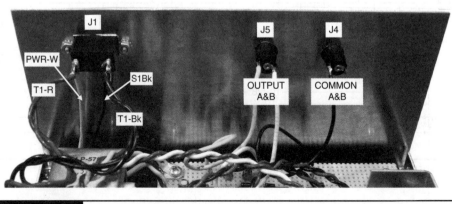

Figure 10-39 Controller rear-panel components.

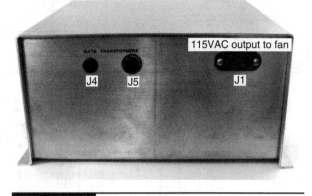

Figure 10-42 Controller box rear panel.

J4: gate transformer return jack to accept the gate transformer return plug from the Full Bridge (see Fig. 10-23).

J5: gate transformer input jack to accept the gate transformer input plug from the Full Bridge (see Fig. 10-23).

3. Interrupter Assembly

Figure 10-43 shows the interrupter with timer, modulator, and audio interface.

Notes on Fig. 10-43 Schematic

Set R17 for 7 V.

Note the 0.1-W 50-V multilayer capacitors across pins 1 and 8 of IC1, IC2, IC3, and IC4—these are C21, C22, C23, and C24.

Fabricate the Interrupter Box by bending and cutting a 1/16-inch aluminum sheet to the dimensions shown in Fig. 10-44 (or plastic may be used for the cover). Alternately, a ready-made metal box (such as an electrical box) may be used with similar or larger dimensions.

Then solder the circuitry onto a perf board (Fig. 10-45).

Interrupter Circuitry Wiring Connections

(All wires except two connect to the interrupter front panel, as shown in Fig. 10-47 and labeled there.)

Figure 10-40 Controller box internal layout and wiring

Controller Rear-Panel Jacks

J1: accepts standard 115-Vac plug to power the Full Bridge cooling fan (see Figs. 10-22 and 10-23).

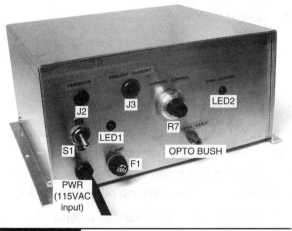

Figure 10-41 Controller box front panel.

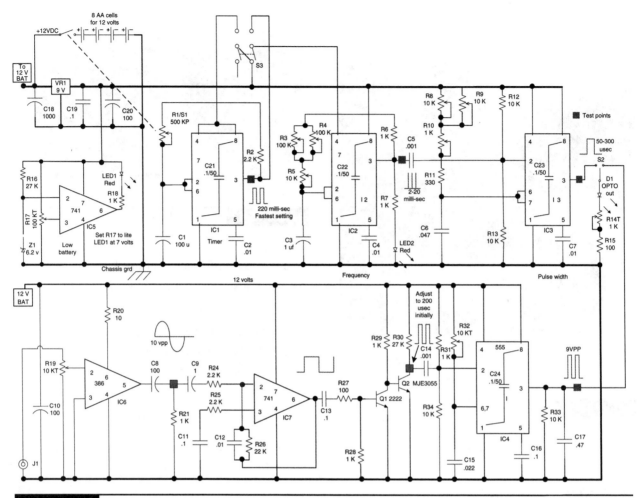

Figure 10-43   Interrupter with timer, modulator, and audio interface.

BATBk: black wire going to battery cradle jack (see Fig. 10-48).

OPTICAL: optical cable going to rear plate (see Fig. 10-48)

Figure 10-46 shows the interrupter circuitry component connections.

Attach and wire the front-panel components (Figure 10-47).

Interrupter Front-Panel Wiring Connections

(All wires except Ti5 connect to the interrupter circuit board, as shown in Fig. 10-45 and labeled there.)

Ti5 (BATRd): red wire going to battery cradle jack (see Fig. 10-48).

Also note where the dial posts are soldered together (see Fig. 10-47):

PULSE WIDTH: There is a trim pot between posts 2 and 3 as shown, and posts 3 and 4 are soldered together.

FREQUENCY: Posts 2 and 3 are soldered together, and there is a trim pot between posts 3 and 4 as shown

TIMER: Posts 2 and 3 are soldered together.

Next, cut an insulating sheet from 1/16-inch plastic to fit across the bottom of the Interrupter Box, place the perf board on top, and drill holes to secure it in place with nuts and bolts (Fig. 10-48).

Label the front panel to indicate control functions (Fig. 10-49).

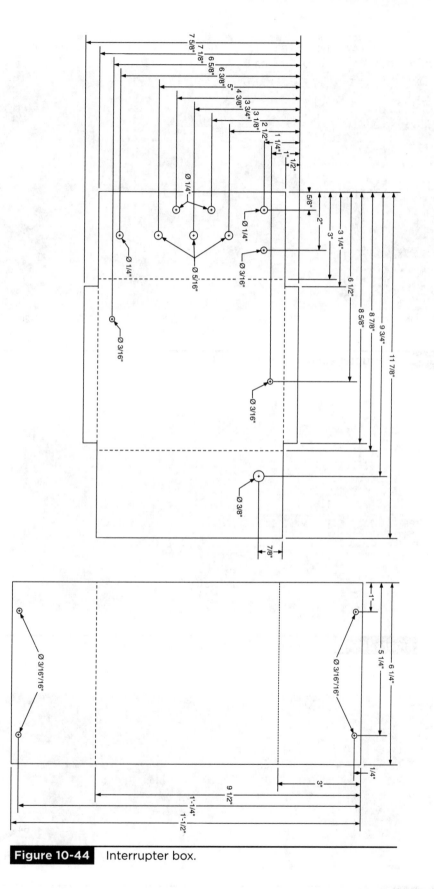

Figure 10-44   Interrupter box.

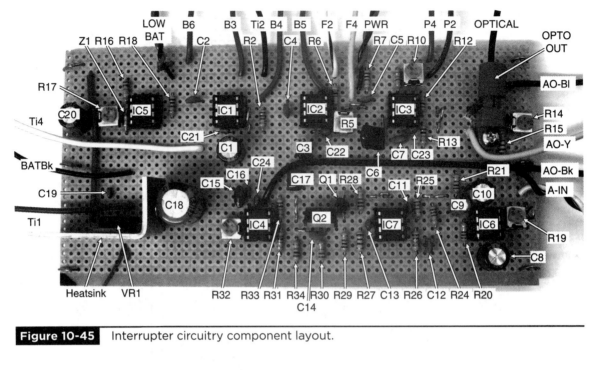

Figure 10-45 Interrupter circuitry component layout.

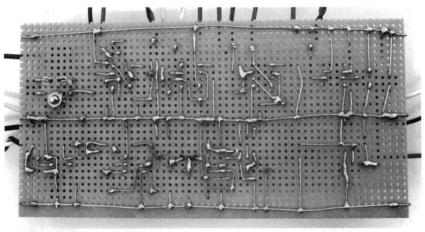

Figure 10-46 Interrupter circuitry component connections.

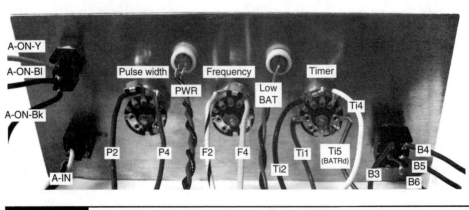

Figure 10-47 Interrupter front-panel component wiring.

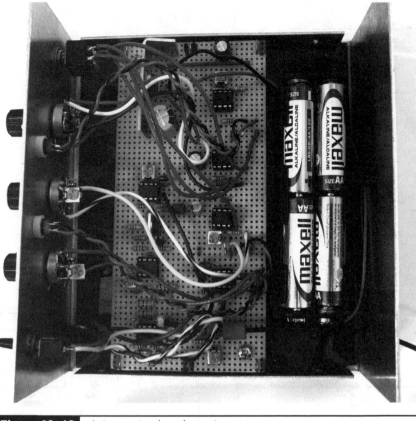

Figure 10-48 Interrupter box layout.

4. Power Supply Assembly

On this power supply (Figs. 10-50 and 10-55), we used a 50-microamp meter for Vdc out, which required R6, R7, and R8 to calibrate the meter to read 0 to 500 Vdc.

Cut a sheet of 1/16-inch plastic to approximately 8½ × 4¼ inches, drilling the holes as required to fit the component dimensions (Fig. 10-52). Note the four mounting holes.

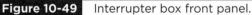

Figure 10-49 Interrupter box front panel.

Solder the components and wiring onto the sheet as shown in Figs. 10-53 and 10-54.

Then wire the components to the ac input power, light-emitting diode (LED), fuse, voltmeter, ammeter, and output power (Fig. 10-55). The Voltmeter is actually a 50-microamp ammeter, but it is wired so that it reads voltage ×10. If desired, for accuracy, it is an easy task to remove the meter's cover with a small screwdriver and place a sticker to cover the "μA" that reads "V × 10" or similar.

The variac is optional but recommended. The variac itself is difficult to build and should be purchased and then wired to the power supply as shown in the schematic in Fig. 10-50 and photographs in Figs. 10-56 and 10-57.

When finished, mount the variac and power supply next to each other on a board (see Figs. 10-56 and 10-57). (The power supply cover here has been painted red to match the purchased variac.)

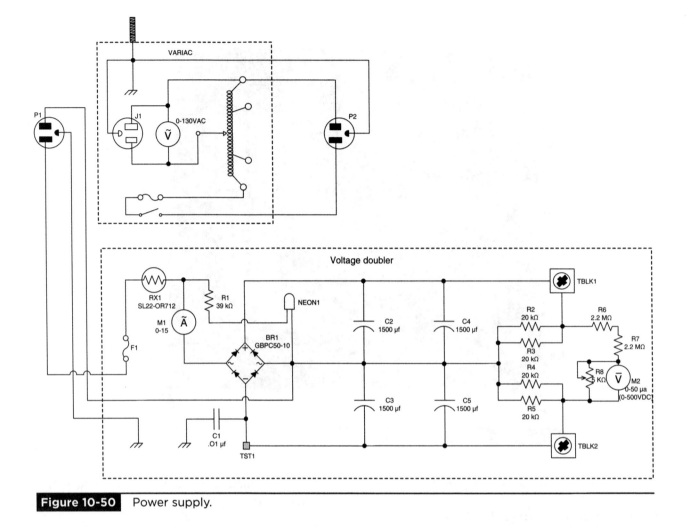

Figure 10-50 Power supply.

5. Coil/Chassis Assembly

5a. Coil/Chassis Assembly: Chassis

Fabricate the chassis top plate as shown in Fig. 10-58 (we use transparent Lexan for a nice visual effect). Dashed lines show the locations of the primary coil, secondary coil, and transformer bracket. Fabricate the bottom chassis the same (½-inch plywood is fine), but with only the four corner holes drilled. The three holes around the larger outer ring (which is where the primary coil will be) are best drilled *after* the primary coil has been made (see Fig. 10-68), so they can be placed according to the coil's actual size.

After the base plates are cut and the four corners drilled, assemble the four base legs as shown Fig. 10-59. It is okay to use a steel threaded

rod instead of brass for these legs because they are far enough away from the primary and secondary to not heat up from the magnetic field oscillation.

Then secure each threaded rod through a corner hole in the base bottom with a nut and washer holding the thin top plastic cap that holds the plastic tube in place and another nut and washer holding the thick bottom polyvinyl chloride (PVC) cap that acts as a foot for the base (Fig. 10-60). Don't snug these up yet because you probably will need to adjust the height of the threaded rod such that it will extend the right amount past the base top plate, strike ring bracket, and washer while still having enough thread left to secure the top nut.

Next, make the Strike Ring by bending copper tubing into a 19-inch-diameter circle and then

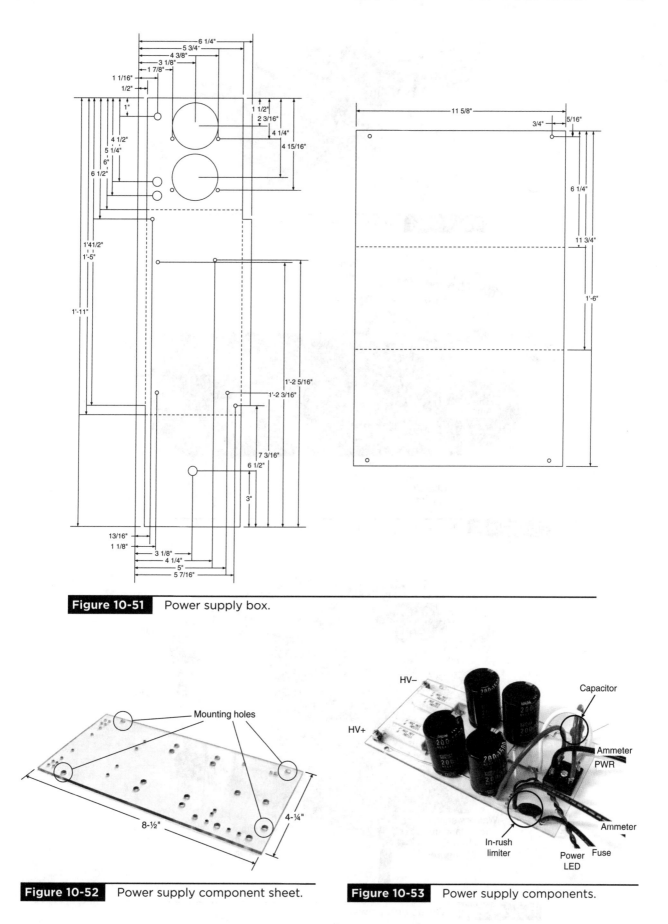

Figure 10-51   Power supply box.

Figure 10-52   Power supply component sheet.

Figure 10-53   Power supply components.

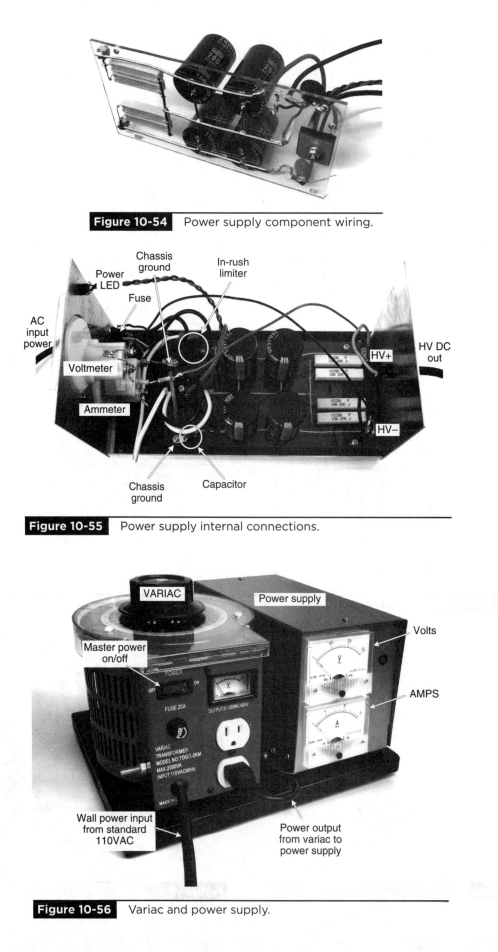

Figure 10-54 Power supply component wiring.

Figure 10-55 Power supply internal connections.

Figure 10-56 Variac and power supply.

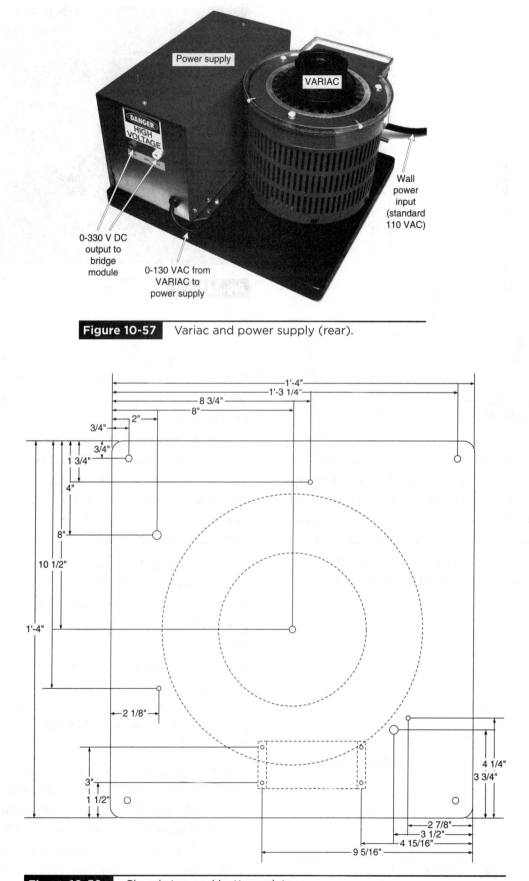

Figure 10-57 Variac and power supply (rear).

Figure 10-58 Chassis top and bottom plates.

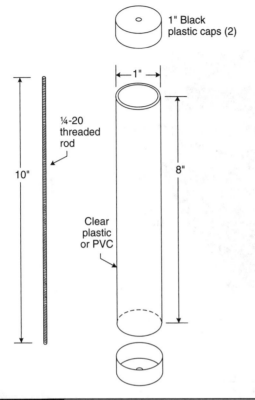

1" Black
plastic caps (2)

¼-20
threaded
rod

10"

1"

8"

Clear
plastic
or PVC

Figure 10-59 Base legs fabrication.

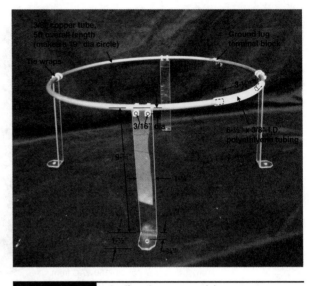

Figure 10-61 Strike ring assembly.

cutting and bending four plastic strike ring
brackets (Fig. 10-61). Copper tube will kink if
bent too sharply, although a 19-inch-diameter
circle should be large enough that kinking won't
occur. If it does, though, then fill a length of
copper tube with water, putting some rubber
caps or something similar on the ends to stop

leakage, and place the tubing in a freezer (a large
"basement" freezer will be required) until the
water freezes. The copper tubing now may be
bent into a tight-diameter circle because the solid
ice core prevents kinking. Once the ice melts,
pour the water out. If you don't have access to
a large enough freezer, you also can pack the
copper tubing with sand. This is not as effective
as a solid-ice core but may help enough to
mitigate the kinking if it's a problem.

Leave about a 3½-inch gap in the strike ring to
keep it from becoming a closed loop, which is then
near enough to the primary and secondary coils that
it would become an active part of this circuit and
cause a parasitic voltage drain between the primary
and secondary coils. The gap is there to prevent
this. A piece of plastic tubing across the gap gives
the strike ring the look of completion and also
prevents anything (such as clothing) from snagging
the ring and pulling on the coil.

The strike ring brackets are made from strips
of plastic bent and drilled to the dimensions
shown in Fig. 10-61. Secure the strike ring to
the top of the brackets with tie wraps. When
this strike ring assembly is ready to be mounted
on the base unit, just slip the feet of the strike

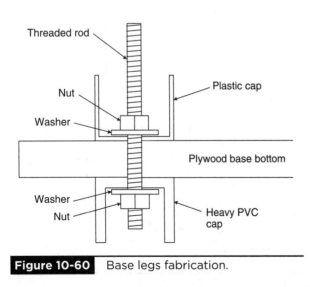

Threaded rod

Nut

Washer

Plastic cap

Plywood base bottom

Washer

Nut

Heavy PVC
cap

Figure 10-60 Base legs fabrication.

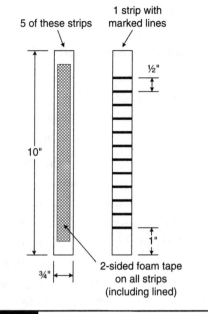

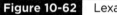

Figure 10-62 Lexan strips.

ring brackets over the exposed threaded rods protruding from the top of the base unit, and cap the rods with a washer and nut.

5b. Coil/Chassis Assembly: Primary Coil

First, cut six Lexan strips ¾ × 10 inches, and mark one of them with lines starting 1 inch from an edge, spaced ½ inch apart. Then cut six pieces of 3M brand ½-inch-wide two-sided foam tape to about 9 inches long, peel off one side (leave the outer cover in place for now), and center on each Lexan strip (Fig. 10-62). Remember that only one strip needs to be marked. Set these aside for now.

Then cut a piece of 1/16-inch Lexan sheet with dimensions 10.5 inches tall by 36 inches long. Then starting 3 inches from one side, mark six vertical lines every 6 inches with a marker; also mark a last line 1 inch from the end of the sheet (Fig. 10-63), which is the strip where the nylon bolts will be going and that will help to align the sheet as it is made into a cylinder.

Next, bend the sheet into a cylinder with 1-inch overlapping ends, and drill 3/16-inch holes for nylon bolts to hold it together. The easiest way to ensure that the holes are aligned is to clamp the overlapping edges between two blocks of wood, which will nearly eliminate warping in the plastic so that when holes are drilled, they will align properly, and the overlapping edges of the Lexan sheet will be as straight and smooth as possible (Fig. 10-64). The cylinder should be about 11.5 inches in diameter.

With the cylinder held together, drill 3/16-inch holes through the overlapping ends, spaced about 1 inch apart. (You may wish to mark a line where the holes will be drilled for a nice straight run.) Then secure the overlapping strips together with nylon nuts and bolts, and release the wood block clamps.

Put the Lexan strips (from Fig. 10-62) on the cylinder along each marked line, centered vertically, held in place with Mylar tape at the top and bottom (Fig. 10-65). The cylinder is now ready to accept the copper coil.

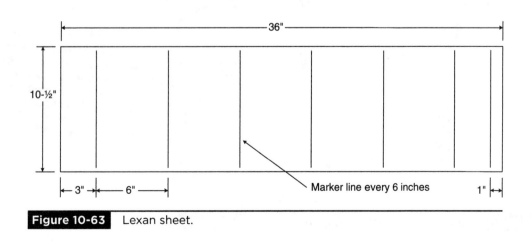

Figure 10-63 Lexan sheet.

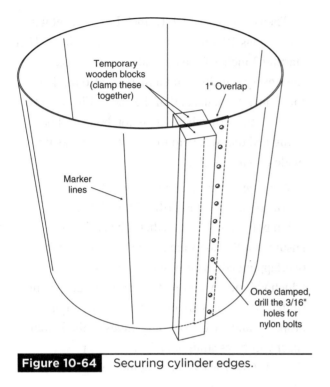

Figure 10-64   Securing cylinder edges.

The copper coil will come wound on a spool about 8 or 10 inches in diameter, so it will need to be opened up by hand to a uniform diameter about ¼ inch larger than your Lexan cylinder (to accommodate the Lexan strips and tape). This will take a little time. Be careful to expand the copper coil in small increments so that nothing is bent too far out of shape. Expand the coil for 12 windings.

With the outer cover still on the tape strips, this is a good time to make sure that your copper coil wraps properly around the cylinder (snug but not too tight is ideal). Make adjustments as needed.

Once you are ready to secure the copper coil around the cylinder, peel off the tape covers, and move the copper coil down the cylinder. This will take some fussing around because there will be 12 lines of coil sticking in six locations, but eventually, step by step, the coil will be maneuvered into place. Take small steps here, "walking" the coil down the tape a little bit at a time to avoid permanently bending any of the copper too far. Use the lines from the single marked Lexan strip to space the coil's windings (Fig. 10-66).

Once the spacing is in place (it need not be perfect but should be as equidistant as reasonably possible), work the looseness out of the coil: Start at the bottom, and secure the trailing end of the copper coil to the cylinder with Mylar tape (bottom

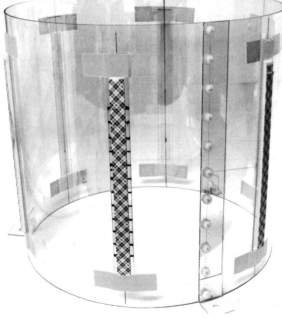

Figure 10-65   Lexan strips attached to cylinder.

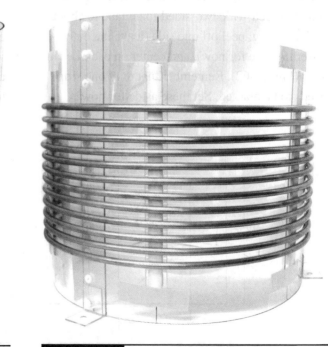

Figure 10-66   Coil spacing around cylinder.

of Fig. 10-67); then push any looseness forward with your fingers to tighten the coil as you go around it. Remember to keep the spacing accurate. This can turn into a bit of a production, and it helps to be an octopus here, but persistence pays off. Once the coil is tightened, secure the top end to the cylinder also with Mylar tape (top of Fig. 10-67).

Verify the coil spacing one last time, and then glue the coils into place along the tape strips (we use a product called Goop, which is a plumbing contact adhesive and sealant). Place a bead of glue on the top and bottom of each winding of the copper coil at each piece of two-sided tape. Be careful not to let the glue go past the tape because you don't want to bond the Lexan strips to the Lexan cylinder (they need to be able to slide around). Let everything dry for about a day.

When the glue is dry the next day, remove the Mylar tape from the copper coil and also from the top and bottom of each plastic strip. Now you have an adjustable-coupling primary coil. The tightness of the coil should hold the plastic strips strongly enough to the cylinder to prevent slipping but also allow the entire coil assembly to slide up and down the cylinder with the application of a little force, which will help to fine-tune the resonance and find the optimal output of the Tesla coil.

For cosmetic reasons, you may want to remove the coil assembly from the cylinder and trim the vertical Lexan strips to about ½ inch above and below the coil. If there is any difficulty in sliding the coil assembly back onto the cylinder, the bottom edges of the vertical strips can be bent outward with a suitable long-nose pliers to make

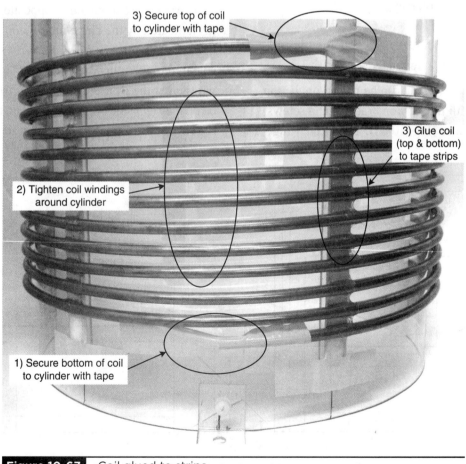

3) Secure top of coil to cylinder with tape

3) Glue coil (top & bottom) to tape strips

2) Tighten coil windings around cylinder

1) Secure bottom of coil to cylinder with tape

Figure 10-67   Coil glued to strips.

Figure 10-68 Base and primary coil (without strike ring).

a slight lip that should help in slipping the coil back onto the Lexan cylinder.

Now the primary coil can be placed on the top of the base with three nylon brackets and screws to hold it in place (Fig. 10-68). The Strike Ring

assembly (from Fig. 10-61) then is placed around the primary coil with the four brackets screwed onto the four legs.

5c. Coil/Chassis Assembly: Current Transformer Bracket

There are two transformers on this bracket: One serves as a feedback loop for the ground wire coming from the Secondary Coil, and the other serves as a current limiter on the HV wire going from the Tank Circuit to the start of the Primary Coil. These transformers are directional, as shown by the circle/arrow icon at the transformer's base (Fig. 10-70). Start by cutting and bending 1/16-inch plastic to form a bracket (Fig. 10-69).

Then mount the transformers to the bracket, solder the wires to the transformers, and glue both the transformers and the wires to the bracket for stability (Figs. 10-70 and 10-71). Note that the transformers are mounted in opposite orientations (as indicated by the circle/arrow icon at the base of each transformer). Also, the wires should be soldered to the transformers such that where the wires enter the Bracket, the circle/arrow icon should be on the *right* side of the transformer, through which the HV wire from the Tank Circuit to the Primary

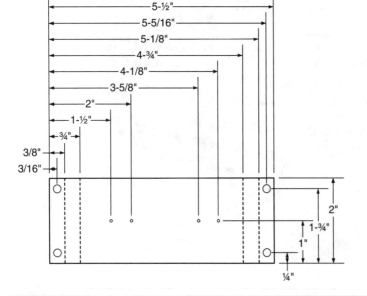

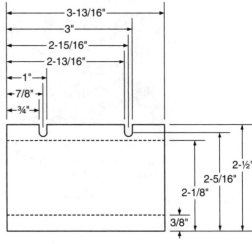

Figure 10-69 Current transformer bracket dimensions.

Figure 10-70   Outer half of bracket with mounted transformers.

Figure 10-72   Inner half of bracket.

Coil will later pass (see Figs. 10-73 and 10-103). The wires should be 14 inches long—too short or too long will bring the entire system out of resonance, so the length is important here. Cap the ends with 2.5-mm plugs (see Fig. 10-73).

The wires are soldered to the transformers such that when looking at the transformer with the circle/arrow icon pointing up, the center wire (the negative wire, usually colored white) is soldered on the right side of the transformer, and the insulating sheath (the grounding/insulating wire, usually a sheath around the center wire) is soldered on the left.

The inner half of the bracket (Fig. 10-72) attaches to the bottom of the outer bracket and covers the exposed wiring; this is simply attached with a thin bead of some glue or plastic epoxy run along the edges.

When finished (Fig. 10-73), this current transformer bracket is attached to the underside of

the base unit on the side where the HV wire comes through the top base plate and attaches to the primary coil (see Fig. 10-103).

5d. Coil/Chassis Assembly: Secondary Coil

NOTE   *Use brass and/or copper (nonmagnetic) fittings where indicated. If steel is used, the magnetic coil field generated will heat up the steel to the point where it will melt through the surrounding plastic.*

The Secondary Coil is an involved assembly that requires a lot of work and attention to detail. Read through the following fabrication procedure. If it looks a little daunting, wound coils are available at www.amazing1.com.

Figure 10-71   Transformer wiring.

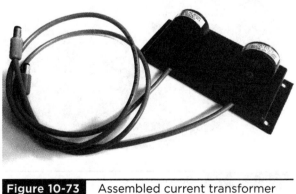

Figure 10-73   Assembled current transformer bracket.

For the top end cap, take some copper sheet (flashing or 1/32-inch-thick copper, whatever you can get) and cut a 6-5/8-inch-diameter circle (or whatever diameter is required to fit inside your secondary-coil tube). Also cut a same-sized circle from ½-inch PVC sheet. Spray one side of the PVC and copper with contact adhesive, and secure them together (two slabs of wood plus two wood clamps work fine here to hold things together until they dry). Once the assembly is dry, drill a hole in the center and tap it for a ¼-20 brass screw about 1½ inches long, and secure this into the cap (Fig. 10-74).

Cut a 3-inch square from ½-inch PVC sheet, and drill a dimple in the center, just deep enough to clear the head of the screw on the top-end cap (Fig. 10-75).

Coat the 3-inch PVC square with silicon adhesive, and secure in place over the screw. Push down tight and snug, and once this has set, further seal around the edges with silicone to ensure that the screw head is as electrically insulated as possible (Fig. 10-76).

For the bottom-end cap, cut a 6-inch-diameter circle from ½-inch PVC sheet, and tap the center for a ¼-20 brass screw. Put a standard aluminum electrical connector on one side (this will be the inside, as per Fig. 10-77) to which the secondary coil winding will later attach (see Fig. 10-95).

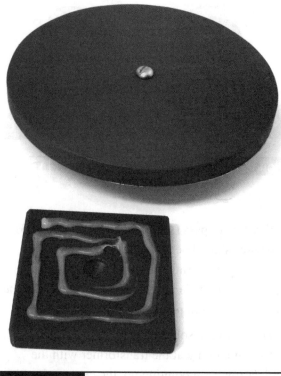

Figure 10-75   Top end cap with screw cover.

Secure with a nut on the other side (what will be the outside, as per Fig. 10-78).

Next, prepare the PVC tube of the secondary coil for the top and bottom end caps. We recommend doing this step before the coil is wound because any mistakes made here will be far less painful without needing to rewind the entire coil.

Figure 10-74   Secondary coil top end cap (outside).

Figure 10-76   Top end cap with cover (inside).

Figure 10-77   Bottom end cap (inside).

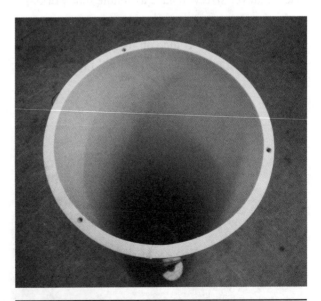

Figure 10-79   Top cap, drilled, on PVC tube.

Start with a 6-5/8-inch outside diameter (OD) PVC tube with 5/16-inch walls (Schedule 40) cut to 40 inches long. On one end (what will be the top), drill three equidistant holes, 1/8 inch diameter, directly down through the top cap near the edges such that they will be centered in the PVC wall. These holes should be 5/32 inch from the edge such that they will center into the 5/8-inch-thick PVC wall of the coil (but adjust if needed according to your material). A drill press is best used here to ensure a nice vertical hole (Fig. 10-79).

Once the holes have been drilled through the top cap, center the cap on the coil, and extend the holes into the PVC coil wall (Fig. 10-80). These should extend about ¾ inch deep into the PVC coil wall.

Countersink the holes in the top cap such that the brass screws will be flush with the top. In our example (Fig. 10-81) we use round-head brass screws, so we countersunk ¼-inch-diameter holes about 3/16 inch deep, but if you

Figure 10-78   Bottom end cap (outside).

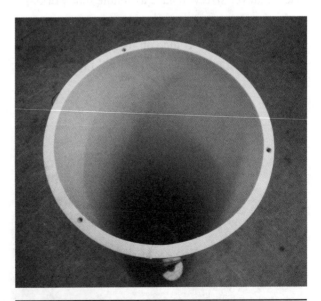

Figure 10-80   Holes drilled into top end of PVC tube.

Figure 10-81 Top cap being attached to PVC tube.

are using angle-head screws, then, of course, use a 45-degree countersink.

The top cap should not yet be attached to the PVC tube because the coil first needs to be wound.

Next, fit the bottom cap.

The bottom cap fits inside the PVC tube instead of covering it, with four equidistant screws to secure the bottom cap in place through the sides. The best way to accomplish this is to first mark and drill all four holes through the PVC tube wall, ½ inch down from the end (Fig. 10-82). We use 8-32 brass screws in these pictures, but you may use whatever size you have available that works—as long as the screws are brass.

Next, lay the PVC tube on its side (this will help the end cap stay in place and not fall into the tube),

and recess the end cap ¼ inch into the PVC tube such that the cap's edges are centered on the four holes. Working on one hole at a time, extend the drill into the cap's edge about ¾ inch deep into the side of the bottom cap. When complete, insert a screw just enough to hold the cap in place, which will make drilling the other holes easier. Do this at each of the four holes, one at a time.

When all four holes are made, mark an alignment arrow on the PVC tube and cap to maintain proper orientation (Fig. 10-83, *upper left, circled*).

Once the end caps have been fitted to the PVC tube, remove them, and next wind the coil.

Winding the coil is a lengthy process requiring patience and attention to detail. We recommend building a simple "lathe" device to hold the coil steady while it is spun. The design of this need not be overly complex because it is simply an axle around which the coil is turned. A rod about 1 or 1½ inches in diameter and 5 ft long (metal or wood) is required, as well as two end caps to hold the coil in place, each end cap consisting of an outer disk larger than the coil's diameter (or about 8 inches in diameter) and an inner disk that fits snugly inside the coil and holds it in place (should be the inside diameter of the coil tube) (Fig. 10-84).

The caps are then snugged up to the coil and secured in place to the axle (we have found that a 1-inch-diameter barbell with circular weight

Figure 10-82 Bottom cap securing holes

Figure 10-83 Bottom cap placement.

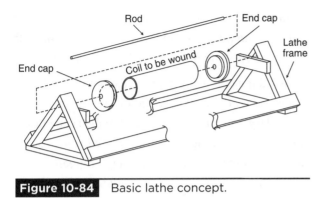

Figure 10-84 Basic lathe concept.

Figure 10-86 Assembled lathe with magnetic wire spool and rod.

clamps works perfectly (Fig. 10-85); alternately, some strong wood clamps or C-clamps may be used or even a couple temporary screws to hold the end caps in place if you're using a wooden dowel—whatever works, but the end caps should be held firmly in place against the PVC tube.

Then another dowel or rod is attached, supported across the A-frame and parallel to the rod holding the PVC tube, to hold the magnetic wire spool as it unwinds onto the PVC tube. The magnetic wire spool can be slid along as progress is made every couple of inches down the PVC tube (Fig. 10-86).

Our shop lathe has an electric motor with a foot-pedal control. Yours need not be so involved if you are just winding one coil, although you may find that an electric motor will help if you are winding this by yourself, depending on how much time you want to spend setting up the electric motor versus winding the coil by hand. Note that winding this coil is an involved project and requires a fair amount of work, and there's no way around this if you're going to do

the work yourself. If you're familiar with motors and chains and sprockets, go ahead and attach an electric motor to spin the lathe. If not, you'll probably overall save time just winding by hand.

Mark a clear indication on the coil, such as colored tape in line with the starting hole (see Fig. 10-89), so that there is a reference for one turn. You will need to count out 1410 turns, so have a pencil and paper handy to track your progress.

With your lathe built and the tube mounted, use one hand to turn the coil and the other to guide the magnetic wire. Or get a helper to steadily spin the lathe while you align the wire. The wire must be kept taught as the coil is wound, but not overly tight. We use a plastic guide to line up the wire as the coil turns and keep the wire taught with the other hand as the wire feeds.

The Plastic Wire Guide is cut from a strip of ¼-inch-thick plastic. Indents for fingers are cut on one side, and a rounded contour is created for the palm on the other side. We then use a heat gun to give it a curve for further ergonomics, or this also can be done by laying the plastic on a metal pipe and putting this in an oven at no more than 200°F (and watching the bending closely so that the

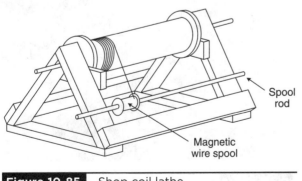

Figure 10-85 Shop coil lathe.

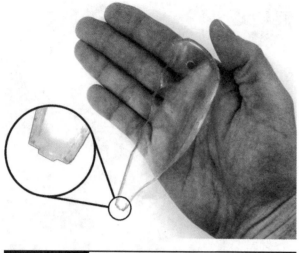

Figure 10-87 Plastic wire guide.

plastic does not melt!). Edges are then smoothed with a file. Pay special attention to the tip because this is what guides and aligns the wire. It would be best to fabricate the tip first and make sure this is correct before putting time into the rest of the design. The tip is basically just an inverted notch at the depth of the magnetic wire that can be formed with a clamp and file (Fig. 10-87).

We set our electric motor to spin the coil at about 1 revolution per second. The wire guide then can be held with one hand while the other hand keeps the wire taught as it feeds from the spool, which gives the wire a nice, clean, smooth winding around the PVC tube (Fig. 10-88).

In either case, when you need to take a break from winding, simply secure the wire to the PVC coil with painter's tape to temporarily hold it in place.

Wind the coil using a 2500-ft spool of No. 24 magnetic wire, in whatever color you prefer (the secondary coil has a large effect on the aesthetics, so choose according to your desire, whether you want a coil with black, red, golden, or green wire—there are many colors available, each of them yielding a unique result). To secure the starting point of the wire winding (which will be the bottom of the coil), either drill one hole large enough to accommodate the No. 24 magnetic wire at 3 inches from the bottom end of the PVC tube, and then pass about 2 ft of wire from the outside into the tube, and secure it in place with superglue. Or drill a row of three holes about ½ inch apart (again, 3 inches from the bottom end of the PVC tube), and thread the magnet wire down one hole, up through the center, and down again through the last remaining hole, again with about 2 ft of wire going into the PVC tube, and this will provide enough friction to keep the wire from slipping as you start the winding. Figure 10-89 shows three holes drilled, but the wire was threaded through only one and held in place with superglue.

Theoretically, the direction of the primary and secondary windings shouldn't matter, but we have them

Figure 10-88 Winding the coil.

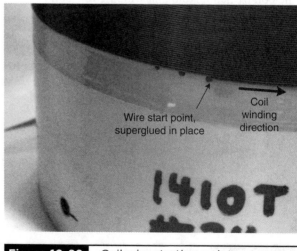

Wire start point, superglued in place

Coil winding direction

Figure 10-89 Coil wire starting point.

both going in the same direction, and that works for us: We start at the bottom and wind counterclockwise as the coils spiral upward (Fig. 10-89).

Once the coil has been fully wound with 1410 turns, the coil winding should come to an end about 6 inches from the top of the PVC tube. Secure the wire a few inches past the 1410th turn with painter's tape; then dab a spot of superglue right at the point of the 1410th turn and let it dry. Then cut the magnetic wire, giving it about 2 ft of slack past the 1410th turn. Once the superglue has dried, remove the tape, and angle the wire gently up and away from the superglued point, and temporarily hold it to the PVC tube with some painter's tape (Fig. 10-90).

Then wrap some painter's tape around the PVC tube about 1 inch beyond the start and stop lines of the magnetic wire (Figs. 10-91 and 10-92) to prepare it for being varnished.

Spin the coil while applying a decent coat of varnish/shellac with a good-quality brush (either foam or bristle) while heating from below (the heater can be seen most clearly in Fig. 10-86). This is where an electric motor helps. The heater is not necessary, but it cures the varnish quicker for less time between coats, especially if you are applying heavier coats, as we do. (The coats need not be

Figure 10-91   Preparing coil for shellac.

excessively heavy but about "candy apple" thickness when applied, which will thin as it hardens.) Let the electric motor spin the coil slowly as the varnish hardens to ensure a nice, even distribution. If you don't have an electric motor, then you will want to

Figure 10-90   Temporarily securing the top of the coil winding.

Figure 10-92   Magnetic wire running to top cap.

spin the coil slowly by hand for the first 5 minutes and then give it a turn every couple minutes for the next 15 minutes. A little extra time on attention to detail at this point will yield a gorgeous finish that you and others will appreciate for decades. Let the varnish dry completely before adding another coat. We apply five coats total, each at about "candy apple" thickness.

Once the last coat of varnish has dried, remove the coil from the lathe, and apply the top and bottom caps. For the top cap, first screw the cap in place to the top of the PVC tube; then run the magnetic wire at about a 45-degree slope up the remaining length of PVC tube, bending it over the copper and holding it in place with tape (see Fig. 10-92).

Then solder the wire to the copper, and cut the excess wire (Fig. 10-93). Finally, coat the wire running along the side of the PVC tube with a layer or two of clear Goop plumbing contact adhesive and sealant to help hold it in place and reduce its chance of snagging on something. If any of the wire is broken, it can be spliced back together again, but of course it's preferable that it never breaks in the first place.

When finished, wipe clear any markings with isopropyl alcohol to yield a clean, good-looking top of your secondary coil (Fig. 10-94).

Figure 10-94 Finished top of secondary coil.

For the bottom cap, cut the magnetic wire to about 9 inches length (long enough that there is enough play to move freely about, but not so long that the wire is prone to bunch up and cross over itself), and attach it to the electrical connector on the inside of the bottom cap (Fig. 10-95).

Figure 10-93

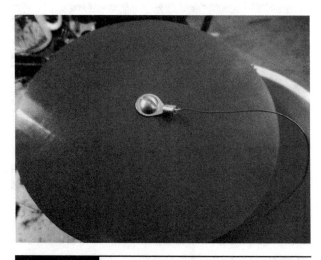

Figure 10-95 Magnetic wire connected to inside of bottom cap.

Figure 10-96 Silicons around inside of bottom cap.

Lay the PVC tube sideways (so that the bottom cap won't fall into it), and insert the bottom cap. As you insert it, turn the cap such that the magnetic wire spirals complementary to how it spirals up the PVC tube. It doesn't need to be a tight curve, just a gentle quarter turn or so. Match the alignment mark (Fig. 10-96), and secure the bottom cap in place with the side screws. Then apply a bead of silicone around the inner lip to reduce the chance of any HV leaks.

Figure 10-97 shows the completed solid-state Tesla coil with all component interwiring.

The wiring overview of all the subassemblies is also shown in the block diagram of Fig. 10-98.

Faraday Cage Assembly

Shielding of your Tesla coil may be necessary depending on the strength of its RF output. See Chap. 20 for information on how to build a Faraday cage.

Your Tesla coil is a resonant high-frequency transformer as well as a highly inefficient radio transmitter. As far as we know, after checking with the Federal Communications Commission (FCC), there has never been an interference report due to the use of these devices.

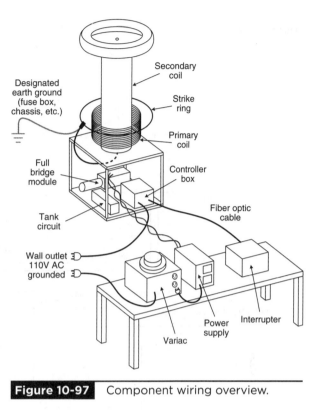

Figure 10-97 Component wiring overview.

Tesla coils, even though with wound secondaries that approximate quarter wavelengths of the operating frequencies, are very poor RF (radiofrequency) radiators. However, some coils can produce radiated emissions in excess of 50 µV/m beyond the legal distance of 3 m. The problem usually can be solved with a reasonably easy-to-construct Faraday cage, as shown in Chap. 20.

Another issue is conducted emissions that can damage sensitive electronic equipment by entrance through shared 115-Vac wiring. This usually can be resolved by installing an appropriate line filter directly on the Faraday cage. Dedicated grounding now must be done properly to avoid this problem.

QUASAR60 Instructions

Figure 10-99 shows the Tesla coil in action.

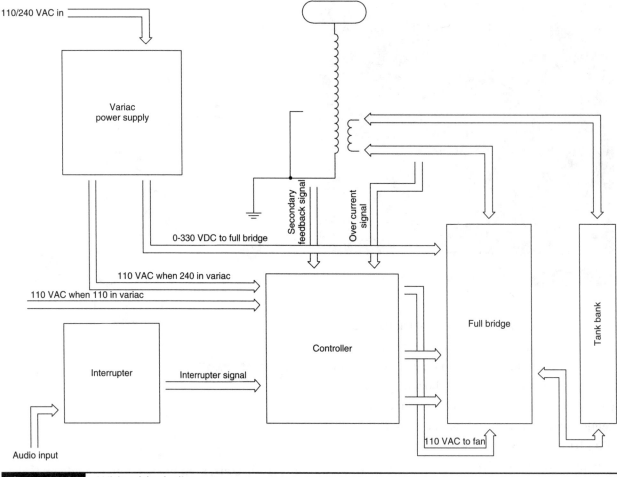

Figure 10-98   Wiring block diagram.

Our experimentation with and testing of Tesla coils are performed in a large, specially made shielded metal building. This building is made with a laminated metal screen floor bonded to the sides of the building and provides almost 100 percent shielding. If you are operating in a congested area or there is sensitive electronic equipment being used nearby you may have to build the Faraday cage.

NOTE   WARNING! SSTDRs REQUIRE A MORE EXTENSIVE ABILITY TO BUILD AND ADJUST THAN THE SIMPLER SPARK-GAP VERSIONS! SMALL OVERSIGHTS CAN CAUSE CATASTROPHIC FAILURE! THIS DEVICE CAN CAUSE DAMAGE TO DIGITAL DEVICES SUCH AS CELL PHONES AND COMPUTERS! DON'T STAND TOO CLOSE TO COIL!

Setup

- Be sure not to put tension on wires as you set up.

- Set the base of unit (with full bridge, tank circuit, and control box) on a workbench or table, and the interrupter and power supply to a different table, then connect a fiber optic cable from interrupter to control box, and all other wiring connections, but do NOT plug into wall power yet (Fig. 10-97, 10-100).

- Attach strike ring (Fig. 10-101) to top of base with acorn nuts (gap in ring with tube goes to rear).

- Attach primary coil to base (seam of tube in back) with nylon screws and nuts. Attach

QUASAR60 instructions

Figure 10-99 Tesla coil in action.

Figure 10-101 Strike ring.

Figure 10-100 Working setup.

Figure 10-102 Primary coil attachments.

Figure 10-103   Secondary coil and strike ring connections.

heavy red wire to lug at base of coil, and clip the tap lead to top of coil (this can be adjusted for tuning and best output; see Fig. 10-102).

- Insert brass bolt on bottom of secondary coil (end with recessed cap).

  - Insert into center hole on base, then attach green ground wire/lug to secondary coil's brass bolt with a brass nut (other end of wire goes through secondary current transformer, and then clips on to strike ring lug, as per Fig. 10-3).

- Place a toroid on top of the secondary coil, and secure with a brass nut.

  - Optionally, screw a brass spine to top of coil above the toroid (this is the breakout point) (Fig. 10-104).

- Remove the cover from interrupter, and insert eight AA batteries into the holder. Reinstall the cover.

- Connect wire from the interrupter to keyboard. (Cable is made to connect in only one direction. Unit will not work if connected improperly.)

CAUTION

- Do not plug controller into variac.

- Always watch ammeter current; 10 to 12 amps is nominal, especially when using keyboard input. (The higher notes draw more current, so adjust the variac accordingly when in audio mode).

- Always let the voltage go down before making any adjustments. (Turn the variac down and

Figure 10-104   Toroid attachment.

let the interrupter, in normal mode, not audio, drain the stored energy.)

■ When adjusting the primary tap, the lower you go (fewer turns), the higher the current rises.

■ During initial setup, set Current Control knob on controller to 0 fully counterclockwise.

■ Turn up the variac in small increments. Adjust (turn up) current control in small increments until output from the toroid is seen; this usually happens around 1.20 on dial. (Current Control LED "red" on controller should still flicker when running; if light doesn't flicker, you will be in danger of damaging components.)

Operation

1. Make sure that all wires are connected properly and that the units are OFF.

2. Plug Controller into 115-Vac outlet.

3. Plug Power Supply into variac and variac into 115-Vac outlet.

4. If you have already done the initial setup, skip to step 12.

5. Set Over Current knob to ZERO.

6. Turn Audio switch down, Frequency knob to the middle, and Pulse Width knob to the middle. Turn Timer knob ON (green light should be ON; if red light is on, replace batteries).

7. Turn Controller ON (green LED should stay ON; red LED should turn ON then turn OFF).

8. Turn Variac to ZERO and power ON.

9. Turn Variac up until you see the Overcurrent light stay ON.

10. Turn Variac down to ZERO and OFF. Wait for voltage to discharge to 20 Vdc on Power Supply. Increase Overcurrent by 10 on inside dial of Current Control on Control Box. (Setting Interrupter to higher frequency will speed up the discharge process.)

11. Repeat steps 9 and 10 until you see Overcurrent light flicker with good output from the Tesla Coil. (Make sure that current as shown on the Power Supply meter does not exceed 13 amps, do step 10 once more.)

12. Turn keyboard on and set to the cleanest sinusoidal tone you can find, as the Interrupter has difficulty translating complex waveforms for use with the Tesla coil.)

CAUTION The current during audio is much greater than during normal operation. Be sure to watch the Power Supply current meter and adjust Overcurrent accordingly. Maximum output during audio may be less than normal operation.

13. Set Interrupter to Audio ON.

14. Turn ON Variac and turn UP while pressing a single midrange note on a keyboard. Do this in 1 second ON, 1 second OFF intervals. (Pay close attention to Power Supply current.)

15. Unit is now ready for operation.

16. To turn unit OFF, set Variac to zero, and let the Interrupter run until voltage reaches 20 Vdc on the Power Supply meter. Turn the Controller OFF, Turn the Interrupter OFF.

Ref. No.	Quantity	Description	Part #
TABLE 10-1	Full Bridge/Tank Circuit Parts List*		
		QUASAR60 Full Bridge	
R7–14	8	10-Ω 1-W film resistor (BR, BLK, BLK)	
C13, 14	2	1-µF 600-V polypropylene	1u/600VP
C15		6300-µF 400-vdc vertical electrolytic	6300u/400V
D1–4	4	IN5819 19-V shotkey	IN5819
D5–12	8	IN4749 15-V zener	
D13–16	4	1.5KE30CA 1.5-kJ 30-V breakover diode	
D17–20	4	MUR1660 dual fast-recovery diode	MUR1660CTG
D21–24	4	1.5KE400CA 1.5-kJ 400-V breakover diode	
Q1–4	4	HGT1N60N60 600-V 60-amp high-powered fet	HGT1N60N60
P4, 5			
P1			
TBLK1–4			
T1, 2		10-turn twisted pair pri 20t twisted pair sec, core OP42212 (Fig. 10-19)	JL6GATE
Fan		UF-80B11 115 Vac	
Heat Sink		Aluminum C-beam 5-1/2" x 3" x 1-1/2" (Fig. 10-4)	FBRHS
Chassis			
Screw			
Solder lug		Assembled Full Bridge Rectifier (Fig. 10-25) available as a ready-to-use module	FBR
		QUASAR60 Tank Circuit	
R1–6	6	10-MΩ 1-W film resistor (BR, BLK, BLU)	
C1–12	12	0.15-µF 2-kVdc polypropylene	.15u/2KVMPP
		Assembled Tank Circuit (Fig. 10-30) available as a ready-to-use module	QUASAR60TC

*Most parts should be available through electronics or hardware stores, but those more difficult to acquire are listed with a "Part #" and are available through www.amazing1.com if needed.

TABLE 10-2 Control Box Parts List*

Ref. No.	Quantity	Description	Part #
		QUASAR60 Controller	
R1		1-kΩ horizontal trimpot	
R2, 12	2	470-Ω ¼-W film resistor (YEL, PUR, BR)	
R3		15-kΩ ¼-W film resistor (BR, GRN, OR)	
R4		560-Ω ¼-W film resistor (GRN, BLU, BR)	
R5, 6	2	2-Ω 2-W film resistor (RED, BLK, GLD)	
R7		10-kΩ 10-turn precision potentiometer	
R8		100-Ω ¼-W film resistor (BR, BLK, BR)	
R9		56-kΩ ¼-W film resistor (GRN, BLU, OR)	
R10		1-kΩ ¼-W film resistor (BR, BLK, RED)	
R11		100-kΩ ¼-W film resistor (BR, BLK, YEL)	
C1, 3		100-μF 25-V electrolytic vertical mount capacitor	
C2, 4, 5, 10, 11	5	0.1-μF 50-V small blue polyester capacitor	
C6		0.001-μF 25-V ceramic disk	
C7		1-μF 25-V electrolytic vertical mount capacitor	
C8		0.0001-μF 25-V ceramic disk	
C9, 12, 15, 17	4	10-μF 25-V electrolytic vertical mount capacitor	
C13, 16, 18, 20, 22	5	0.1-μF 50-V small blue polyester capacitor	
C14		4.7-μF vertical electrolytic	
C19, 21, 23, 24	4	10-μF 25-V electrolytic vertical mount capacitor	
D1		1.5KE7.5CA 7.5-V tranient voltage suppression 1500 W	
D2		1.5KE18CA 18-V tranient voltage suppression 1500 W	
D3		IF-D92 opto in	
D4, 6, 8, 9	4	IN5819 19-V shotkey diode	IN5819
D5, 7	2	IN5337 1-W zener doide	
D10–13,14–17	4	IN5819 19-V shotkey diode	In5819
LED1		Green indicator led	
LED2		Red indicator led	
IC1, 2		SN7414N HEX inverting schmitt trigger 14-pin dip	
IC3		SN74HC74 JK flip flop 14-pin dip	
IC4		SN74HC08 and gate 14-pin dip	
IC5		LM311 comparator 8-pin dip	
IC6		LM555 timer 8-pin dip	
IC7, 8	2	UCC37322P high-speed low-side mosfet driver with enabled 8-pin dip	UCC37322
IC9, 10	2	UCC37321P high-speed low-side mosfet driver with enabled 8-pin dip	UCC37321
T1		LP-575 120 V/24 V	
T2,3	2	Triad CST206-1A	
S1		SPST	
F1		10-AMP FUSE HOLDER, 2-AMP FUSE	

*Most parts should be available through electronics or hardware stores, but those more difficult to acquire are listed with a "Part #" and are available through www.amazing1.com if needed.

TABLE 10-2 Control Box Parts List (*Continued*)

Ref. No.	Quantity	Description	Part #
		QUASAR60 Controller	
J1		2-prong 115-V jack	
J2–5	4	neon 2.5-V jack	
SOCK8X	6	8-pin IC socket for IC5, 6, 7, 8, 9, 10	
SOCK14X	4	14-pin IC socket for IC1, 2, 3, 4	
Heat sink			
Chassis			
		½-inch cord clamp	
Bushing			
Raiser			
Screw			
Nut			
CH screw			
Opto bush		Bushing for optical cable	HFBR-4505Z
		Optional printed circuit board for Control Module	PCQUASAR60-CB
		Assembled Control Box (Fig. 10-41) available as a ready-to-use module	QUASAR60-CONTROLLER

TABLE 10-3 Interrupter Parts List*			
Ref. No.	Quantity	Description	Part #
		QUASAR60 Interrupter	
R1/S1		500-kΩ 17-mm linear pot w/ switch	
R2, 24, 25	3	2-kΩ ¼-W film resistor (RED, RED, RED)	
R3		100-kΩ 17-mm linear pot	
R4		100-kΩ vertical trimpot	
R5, 32	2	10KΩ horizontal trimpot	
R6, 7, 18, 21, 28	6	1-kΩ ¼-W film resistor (BR, BLK, RED)	
R8		10-kΩ 17-mm linear pot	
R9, 19		10-kΩ vertical trimpot	
R10, 14	2	1-kΩ horizontal trimpot	
R11		330-Ω ¼-W film resistor (OR, OR, RED)	
R12, 13	4	10-kΩ ¼-W film resistor (BR, BLK, OR)	
R15, 27	2	100-Ω ¼-W film resistor (BR, BLK, BR)	
R16, 30	2	27-kΩ ¼-W film resistor (RED, PUR, OR)	
R17		100-kΩ horizontal trimpot	
R20		10-Ω ¼-W film resistor (BR, BLK, BLK	
R26		22-kΩ ¼-W film resistor (RED, RED, OR)	
R29, 31		1-kΩ ¼-W film resistor (BR, BLK. RED)	
R33, 34	3	10-kΩ ¼-W film resistor (BR, BLK, OR)	
C1, 8, 10, 20	4	100-µF 25-V vertical electrolytic	
C2, 4, 7, 12	4	0.01-µF 50-V disk capacitor	
C3, 9	2	1-µF 25-V vertical electrolytic	
C5, 14	2	0.001-µF 50-V disk capacitor	
C6		0.047-µF 50-V polyester vertical mount cap	
C11, 13, 16	3	0.1-µF 50-V multilayer cap	
C17		0.47-µF 50-V metalized polyester	
C18		1000-µF 25-V vertical electralitic	
C19, 21–24	5	0.1-µF 50-V multilayer cap	
D1		IF-E91A opto out	IF-E91A
LED1		Red indicator led	
LED2		Green indicator led	
S2		DPST small 115-vac 3-amp toggle switch	
S3		DPDT small 115-vac 3-amp toggle switch	
Q1		PN2222A 40-V 20-milliamp *npn* general-purpose transistor plastic	
Q2		MJE3055 *NPN* power transistor plastic	
VR1		L7809ACV 9-V regulator	
IC1–4	4	LM555 8-pin timer	
IC5, 7	2	LM741 operational amplifier 8-pin dip	
IC6		LM386 operational amplifier 8-pin dip	

*Most parts should be available through electronics or hardware stores, but those more difficult to acquire are listed with a "Part #" and are available through www.amazing1.com if needed.

Ref. No.	Quantity	Description	Part #
TABLE 10-3 Interrupter Parts List (*Continued*)			
		QUASAR60 Interrupter	
SOCK8X	7	8-pin IC socket for LM555 and LM741	
Bushing		Led bushing	
Opto bush		Bushing for optical cable	HFBR-4505Z
Bat hold		8 AA cell battery holder	
Bat con		Mouser electronics 121-1022/I-GR	
Chassis		Chassis Fabricate Per Fig. 16-44	
Heat sink		2¼ × ¾ × 1/16″ aluminum	
Nut			
Screw			
Nut			
Screw			
		Optional printed circuit board for Interrupter Module	PCQUASAR60-IN
		Assembled Interrupter Box (Fig. 10-49) available as a ready-to-use module	QUASAR60-INTERRUPTER

Ref. No.	Quantity	Description	Part #
TABLE 10-4 Power Supply Parts List*			
		Quasar Power Supply	
R1		39-kΩ ¼-W film resistor (OR, WH, OR)	
R2–5		20-kΩ 10-W power resistor	
R6, 7		2.2 MΩ 1 W film resistor (RED,RED,GRN)	
R8		5-kΩ vertical trimpot	
RX1		In-rush limiter sl22-2r515	
C1		0.01-µf 2-kv disk capacitor	
C2–5		1500-µF 200-V vertical electrolytic	1500u/200V
BR1		GBPC50-10 1000 v 50 amp	RECT50A1KBR
F1		10-amp fuse holder, 15-amp fuse	
NEON1		Neon idicator bulb	
M1		15-amp meter ac line current	MET15AAC
M2		50- µa meter dc overcurrent and other functions	METER50L
TBLK1, 2			
CO1		Three-wire no. 14 power cord	CORD143
	6	solder lugs	
Screws			
Nuts			
	2	½-inch cord clamps	
Chassis		Chassis fabricate per Fig. 16-51	
Base		12½- × 13½- × ¾-inch plywood ¾-inch radius corners	
Variac		Tdgc-2kva, input 110 vac, output 0–130vac 20 amps	VARIAC2KVA
Risers		½-inch risers	
Insulater			
Board			
Lid			
Neon bush			
Wire nut			
		Assembled Power Supply (Fig. 10-56) available as a ready-to-use module	QUASAR60-PSU

*Most parts should be available through electronics or hardware stores, but those more difficult to acquire are listed with a "Part #" and are available through www.amazing1.com if needed.

Ref. No.	Quantity	Description	Part #
TABLE 10-5 Coil/Chassis Parts List*			
		QUASAR60 Coil/Chassis	
		Primary feedback/ over current incloser	
		16- × 16- × ¾-inch plywood	
		16- × 16- × 3/8-inch plastic	
	4	8-inch red 1-inch tube	
	4	10-inch 1/4-20 threaded rod	
		6-inch white schd 40 pvc tube with 1400 turns no. 24 magnet wire (secondary coil)	QUASAR60COIL
		6½-inch-diameter ¾-inch PVC	
		11 turns of ¼-inch copper tube roughly 12 ft for primary coil	
		40- × 10-3/8- × 1/16-inch plastic for primary coil mount	
		5 ft of 3/8-inch copper tube for ground strike	
		TBLK for ground strike morris no. 90712	
		6-inch thick x 12-inch inner diameter x 24-inch outer diameter spun aluminum toroid	TO24

*Most parts should be available through electronics or hardware stores, but those more difficult to acquire are listed with a "Part #" and are available through www.amazing1.com if needed.

Six-Foot Jacob's Ladder

Overview

A spectacular display of traveling plasma-arc discharges for museums, attention-getting advertising, special events, and so on.

Hazards

Uses dangerous high voltages necessary for powering required electronic circuitry. High-voltage handling skills are *positively* required for building this project! Eye protection should be worn when making and testing this device.

Difficulty

Requires *advanced* skills in wiring and soldering, along with *experience* in fabricating sheet metal and plastics and woodworking. Electronic knowledge in the fields of high-voltage (HV) devices should be familiar to the builder. Construction is shown as individual modules, each being built and tested.

Tools

Wiring tools of several sizes, large 100- and 50-W soldering irons, hand and power tools, fabrication equipment, good voltmeter and ammeter, and *low-cost oscilloscope*. Electronic ability to use test equipment such as meters and a scope will help in understanding and completing this project.

You should familiarize yourself with Chap. 21 before building this circuit or any similar circuit without an isolation transformer. We advise that you use the testing-circuit jig shown in Chap. 21 for this project because it verifies dangerous alternating-current (ac) grounds. **Other amenities are powering a virgin circuit starting with a low input voltage and slowly increasing the voltage, noting that any excessive current could be dangerous and totally wipe out your hard work!**

Assembly

The assembly of this large Jacob's ladder (Fig. 11-1) is divided into five parts:

1. Control board, timer, and power supply
2. Transformer

Figure 11-1   Large Jacob's ladder.

3. Full bridge and bridge rectifier

4. Base

5. Chimney and ladder

1. Assembly: Control Board, Timer, and Power Supply

Figure 11-2 shows the control board, timer, and power supply this Jacob's ladder. The circuitry for this Control Board is shown in Fig. 11-3 laid out on a standard perf board, but a printed circuit board (PCB) is also available and can be found on our website at www.amazing1.com (Item

PCJACK60-CB) that will provide an easier build for those who want it.

Before starting the soldering, take a marker and trace out component locations on one side of the perf board and the wiring connections on the other side. Then solder the components and wires to the perf board as shown in Figs. 11-3 and 11-4.

Control Board Wiring Connections

220 Vac IN: connect black wires together and red wires together; these go to the incoming 220-Vac power supply

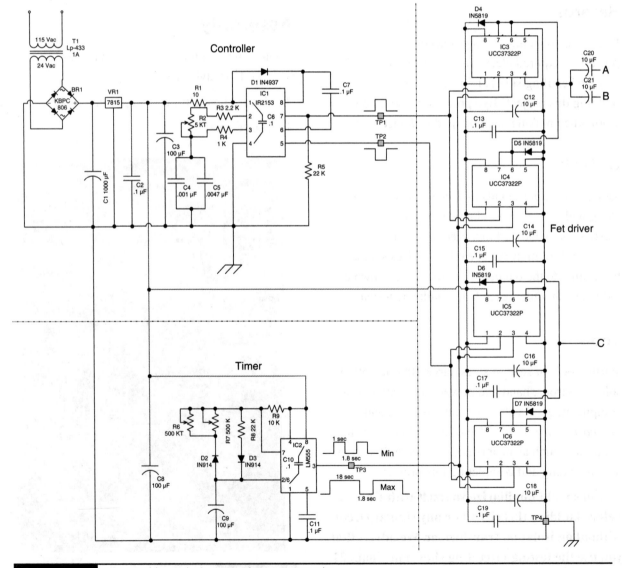

Figure 11-2   Control board, timer, and power supply.

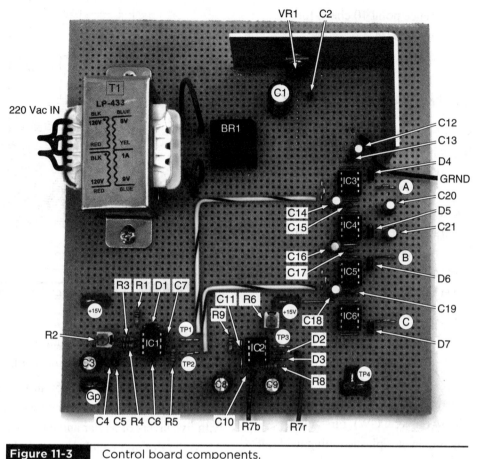

Figure 11-3 Control board components.

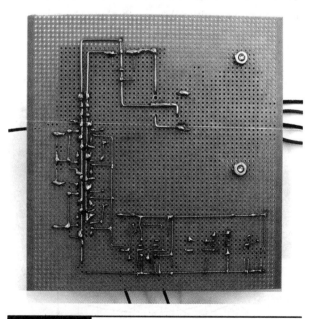

Figure 11-4 Control board wiring connections.

R7b: black wire going to R7 "On Time" adjustment dial (see Fig. 11-39)

R7r: red wire going to R7 "On Time" adjustment dial (see Fig. 11-39)

GRND: black wire going to chassis ground (see Fig. 11-39)

A: connection point for one set of inside gate transformer wires (see Fig. 11-19)

B: connection point for one set of inside gate transformer wires (see Fig. 11-19)

C: connection point for both sets of outside gate transformer wires (see Fig. 11-19)

TP1, 2, 3, 4: test points to read wave output characteristics on scope (TP4 is ground; see schematic in Fig. 11-2)

+15V: 15-V reference point 10 check for correct line voltage on DMM or scope

Gp: ground reference point

2. Assembly: Transformer

First, cut a base from a 4- × 12- × ¼-inch Lexan sheet (or acrylic, polyvinyl chloride, or wood—anything electrically insulating). Then cut two small polyvinyl chloride (PVC) blocks to the same height as the base of the ferrite cores (these are used to balance the outer transformer windings). Epoxy or superglue the ferrite cores to the base, and then superglue the PVC support blocks to the sides of the ferrite cores. Cover the ferrite-core tops with 2-mil Mylar tape (Fig. 11-5), which will create a gap between the top and bottom ferrite cores.

2a. Transformer Secondary Coils

The two outer secondary coils must be wound in opposite directions (CW & CCW). First, cut a section of 1½-inch-diameter plastic tubing to about 2-3/8 inches length (the actual length should be about 1/8 inch less than the gap between the lower and upper ferrite cores to prevent the coils from pushing against the cores and opening up the gap), which will be the inner cylinder around which the coil wire is wound. Then cut a section of 3¼-inch-diameter plastic tubing, which will be the outer covering, to the same 2-3/8-inch length. Repeat inner and outer cylinders for the other secondary coil of the transformer.

Wind No. 24 magnetic wire a total of 700 times around the inner cylinder. Start the winding ½ inch from the edge of the tube to prevent proximity

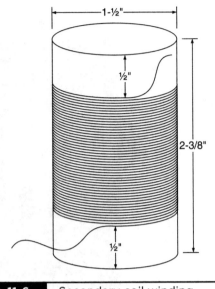

Figure 11-6   Secondary coil winding.

arcing with the ferrite core, and stop winding ½ inch from the other end of the cylinder (Fig. 11-6).

You should be able to get 50 turns over this distance, after which you should wrap the windings with five layers of Mylar and cut a ¼-inch notch in the Mylar through which the magnetic wire is pulled (Fig. 11-7). Making this cut requires attention to detail. It will require either a very fine pair of scissors (such as those found in Swiss Army knives or some other multitools) or a thin and sharp blade

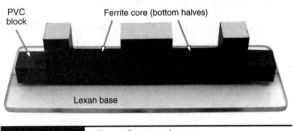

Figure 11-5   Transformer base.

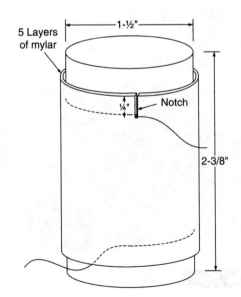

Figure 11-7   Notches in the Mylar.

to give precise control and reduce the chance of cutting or nicking the magnetic wire, such as an X-acto knife or razor blade to score the cut (which may require a few passes).

It is also important to stagger the cut location each time (at both the top and bottom); otherwise, the notches will line up and increase the chance of an arc jumping across the cut locations of the magnetic wire. So every time 50 windings have been made and a notch is to be cut, just go about ¼ or ½ inch past the previous notch (Fig. 11-8). Do this on both the top and bottom sides.

Take the outer plastic 3½-inch cylinder and drill a hole in it near the top, large enough to pass through the HV wire (see Fig. 11-10 for its relative location). Then epoxy a 3½-inch bottom cap onto the cylinder, and let this set before continuing (may require a few hours). Sealing the cap to the outer cylinder will force the air to be drawn up through the inner cylinder when the coil is later put into a vacuum jar, pulling the goop (a two-part silicone elastomer) down into the windings (see Fig. 11-12).

After the cap epoxy has dried, pierce a small hole through the bottom cap (with a needle or small

Figure 11-9   Secondary coil bottom.

nail)—and make sure that this hole in the bottom cap is placed such that the hole in the *top* of the cylinder, for the HV wire (see Fig. 11-10), will align with the top of the coil winding when the coil is seated in the cylinder. With this positioned right, run the coil's magnetic wire through the hole in the bottom cap until the coil is seated, and seal the bottom hole with epoxy (Fig. 11-9). The best way to do this effectively is to either place the coil upside down and then lower the cylinder/cap over the coil while pulling through the magnetic wire, or turn everything on its side and push the coil into the cylinder/cap while pulling through the magnetic wire. The idea is to prevent the magnetic wire from bunching up at the bottom of the coil in a tangle, but also the magnetic wire should not be drawn too tight. The wire should just flow normally from the bottom of the coil through the hole in the end cap.

Next, solder the HV wire to the top of the coil winding. Also put a cap into the inner cylinder (which will prevent the goop from overflowing into it), and puncture it with several holes through which air can escape (Fig. 11-10).

Make another coil, winding it in the opposite direction from the first.

Put a scrap sheet of wood on the bell vacuum base (in case of any spillage), and then place the coils on this wood. Pour goop into a transformer, slowly spreading it evenly around the top. When one transformer is nearly filled, pour goop into the other transformer. The goop will settle between

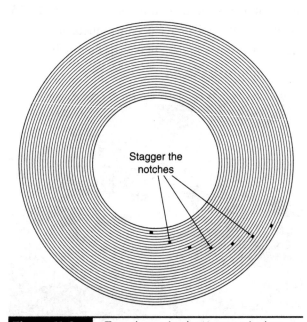

Figure 11-8   Top-down (or bottom-up) view of coil with staggered cuts in the Mylar.

Stagger the notches

Figure 11-10 Wound secondary.

Figure 11-12 Coils in a vacuum.

the windings, so go back and forth between the transformers adding goop as needed (Fig. 11-11).

When the rate of settling begins to slow down, put the cap on the vacuum bell, and turn on the pump to help draw the goop into the windings (Fig. 11-12).

When the goop has been drawn far enough down into the coils such that the tops of the windings start to be seen again, turn off the pump, top off the coils with more goop, and then reseal the jar and turn on the vacuum pump again. You want to fill the coils with goop and avoid any air pockets. This may need to be

repeated a few times until the windings have been completely filled top to bottom with goop. Figure 11-13 shows a partially saturated coil being topped off with goop for another treatment with the vacuum jar.

Once the coils are sealed, let them dry (usually overnight for 8 hours should suffice). Then place them on the ferrite cores; be sure that any plastic housing and hardened goop are trimmed down as needed such that the top halves of the ferrite cores rest completely on the bottom halves. Figure 11-15 shows how the coils have been trimmed, and Fig. 11-16 shows the top halves of the ferrite

Figure 11-11 Pouring the two-part silicone elastomer goop.

Figure 11-13 More goop in the partially filled coils.

coils resting with full contact on the bottom halves.

2b. Transformer Primary Coil

For the inner primary coil, cut and bend a small sheet of 1/16-inch plastic, and fabricate it into a box for the inner coil winding (Fig. 11-14). The dimensions given are approximate; make sure that the box fits snugly around the center block of the ferrite cores (the tighter the better), taking into account the loss in dimensionality when the plastic is bent. First bend the box into its rough four-sided shape. Then bend the top and bottom tabs outward, which will hold the Litz wire in place when it is wrapped around the box. Wrap the box with Mylar tape to hold it together; remember that this is not a structural component and only needs to serve as a guide for the Litz wire.

With the box finished, wind the twisted-strand No. 16 Litz wire 24 times around the box. The wire should start at one bottom corner and wrap 12 times up; cover the wires with five layers of Mylar, and then wrap back 12 turns down, exiting at the adjacent bottom corner. Then wrap this coil again with another five layers of Mylar tape (Fig. 11-15).

Complete the transformer by laying the remaining top two ferrite cores in place (Fig. 11-16). The weight of the cores should be enough to hold the transformer together, but feel free to drill some holes in the base and latch down each half of the transformer with some tie wraps if there is concern that the cores may jostle out of position (if the Jacob's ladder will be moved around, for example).

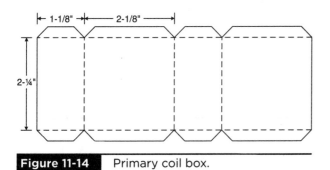

Figure 11-14   Primary coil box.

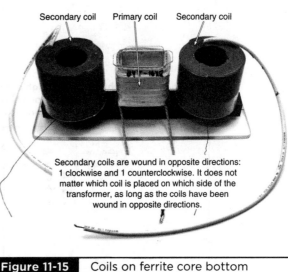

Secondary coils are wound in opposite directions: 1 clockwise and 1 counterclockwise. It does not matter which coil is placed on which side of the transformer, as long as the coils have been wound in opposite directions.

Figure 11-15   Coils on ferrite core bottom halves.

The two HV output wires are then each soldered to a standard electrical ring terminal, insulated with some shrink tubing, and later attached (bolted) to the terminal blocks of the ladder (see Fig. 11-33).

Transformer Wiring Connections

WHV1: high-voltage output wire going to the base electrical block of the ladder rod (see Fig. 11-33).

WHV2: high-voltage output wire going to the base electrical block of the ladder rod (see Figs. 11-33).

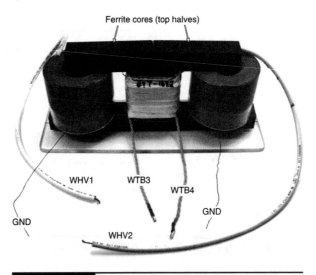

Figure 11-16   Assembled transformer.

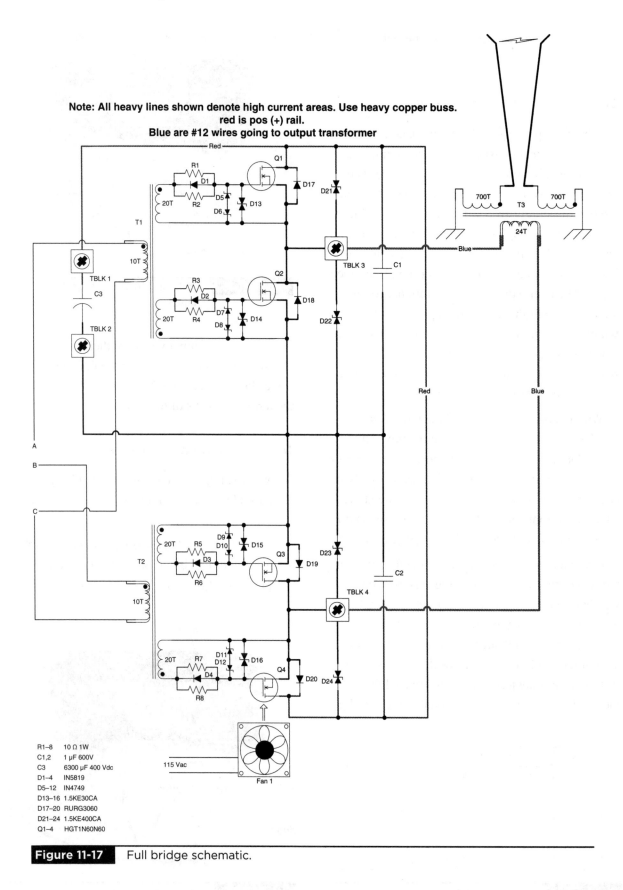

Note: All heavy lines shown denote high current areas. Use heavy copper buss.
red is pos (+) rail.
Blue are #12 wires going to output transformer

R1–8   10 Ω 1W
C1,2   1 µF 600V
C3   6300 µF 400 Vdc
D1–4   IN5819
D5–12   IN4749
D13–16 1.5KE30CA
D17–20 RURG3060
D21–24 1.5KE400CA
Q1–4   HGT1N60N60

Figure 11-17 Full bridge schematic.

WTB3: two-strand Litz wire going to full bridge terminal block TBLK3 (see Fig. 11-18; no polarity, so this may be connected to either TBLK3 or TBLK4).

WTB4: two-strand Litz wire going to full bridge terminal block TBLK4 (see Fig. 11-18; no polarity, so this may be connected to either TBLK3 or TBLK4).

GND: secondary coil winding going to chassis ground.

3a. Assembly: Full Bridge

The Full Bridge schematic (Fig. 11-17) and assembly (Fig. 11-18) are the same as in the QUASAR60 Solid-State Tesla Coil (Figs. 10-3 to 10-25) minus the Tank Circuit.

Full Bridge Contacts

TBLK1: positive (+) rail of ~330-Vdc input going to BR1 (see Fig. 11-21).

TBLK2: negative (−) rail of ~330-Vdc input going to BR2 (see Fig. 11-21).

TBLK3: square-wave HV input/output going to primary coil of transformer T1 (see Fig. 11-16; can be connect to either wire).

TBLK4: square-wave HV input/output going to primary coil of transformer T1 (see Fig. 11-16; can be connect to either wire).

NOTE Do not put jacks on the gate transformer wires (as seen in Fig. 11-18) because these wires are instead soldered directly to the Control Board: the outer wires (Fig. 11-19) are all connected to the same point C in Fig. 11-3, and the inner wires go to points A and B (there is no polarity, so it doesn't matter which inner wires go to point A or point B as long as one set of inner wires go to point A, and the other set of inner wires go to point B).

3b. Assembly: Bridge Rectifier

To make the Bridge Rectifier, start with cutting a 4½- × 3¼-inch aluminum sheet, and drill six 5/32-inch holes through it (Fig. 11-20). Also cut two aluminum strips approximately ½ inch wide by 2-1/8 inches long, and drill a ¼-inch hole at each end. The actual placement of these holes will depend on the dimensions of the dual diodes being used.

Then secure the two dual diodes onto this aluminum sheet with ½-inch 6-32 screws from beneath and lock washers on top. Run the two aluminum strips across the top connections so that the positive (+) terminals contact each other and same with the negative (−). Then attach four electrical terminals to the top contacts, and a smaller electrical terminal beneath one of the lock washers on the dual diode for the ground (Fig. 11-21).

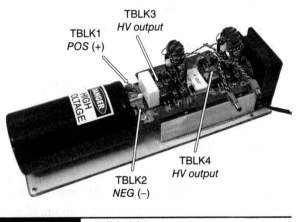

Figure 11-18 Full bridge.

Figure 11-19 Gate transformer wiring.

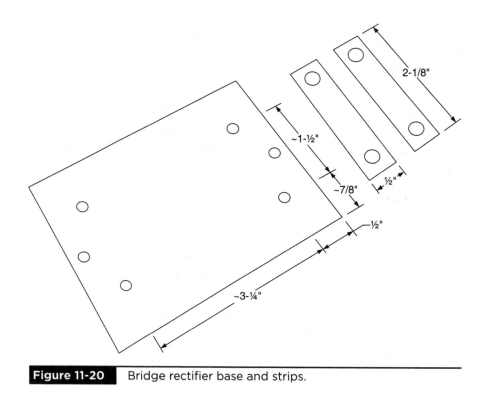

Figure 11-20 Bridge rectifier base and strips.

Bridge Rectifier Contacts

220ACIN: 220-Vac input wire (see Fig. 11-39).

BR1: positive (+) going to TBLK1 connection (see Fig. 11-18).

BR2: negative (−) going to TBLK2 connection (see Fig. 11-18).

GROUND: to chassis ground.

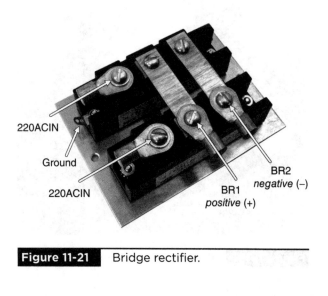

Figure 11-21 Bridge rectifier.

4. Assembly: Base

Four separate sheets of aluminum are used in base fabrication. Start with a ½-inch sheet of plywood approximately 17 × 17 inches square, and then cut a plastic insulating sheet to be placed on top of this and the ½-inch clear acrylic cover that will serve as the top plate (Fig. 11-22).

Then drill holes in the top plate—a center hole 3/8 inch in diameter, eight venting holes about ¼ inch in diameter in a tight group at the center (in about a 3-inch-diameter circle), another eight venting holes wider out (in about an 8-inch-diameter circle but still within the chimney walls), and then two more holes for the terminal blocks that hold the ladder and connect it to the HV transformer outputs about 4 inches apart. The venting holes should be chamfered, if possible; the terminal-block holes should be straight through (Fig. 11-23).

Now cut four sheets of aluminum, and drill and bend as shown in Fig. 11-24. The aluminum sides should be sized to fit around the plywood base.

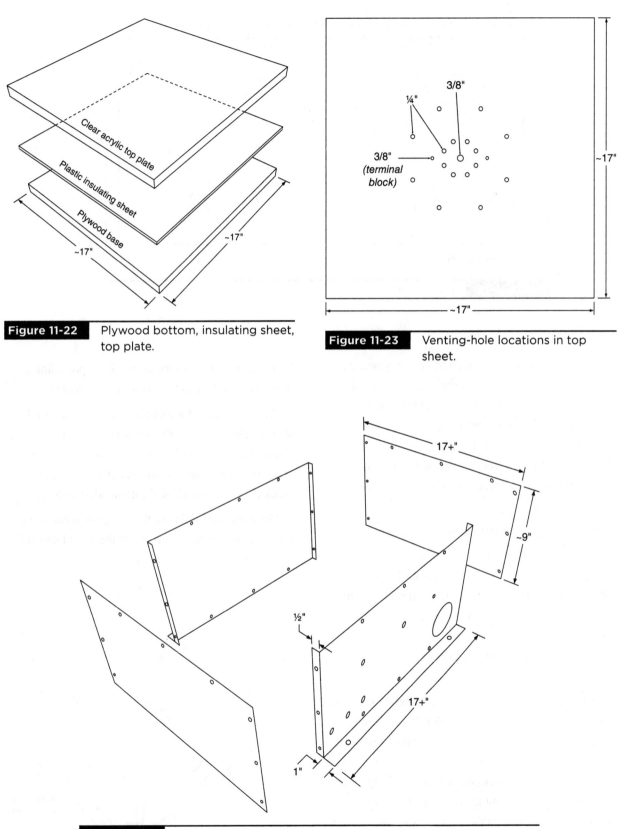

| **Figure 11-22** | Plywood bottom, insulating sheet, top plate. |

| **Figure 11-23** | Venting-hole locations in top sheet. |

| **Figure 11-24** | Aluminum base sides. |

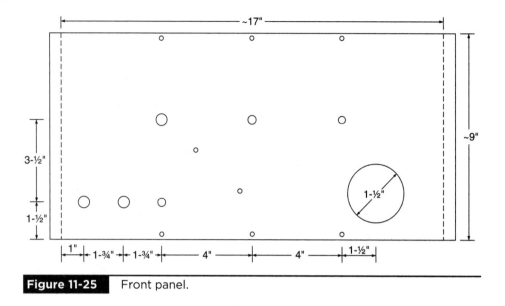

Figure 11-25   Front panel.

The hole locations don't need to be exact because they are just used to hold the base together. Thus they can be drilled as the sides and bottom are assembled and then secured in place with screws and nuts and bolts. A height of about 12 inches should be sufficient to clear all the electronics and leave enough head space for sufficient airflow, but adjust as needed.

The front sheet should have holes drilled for components (Fig. 11-25).

First, secure only the bent sides to the plywood base (note that the flat-sheet sides do not need to be secured to the base, but if they are, it may be desirable to put washers between the sheet and base to prevent the sheet from bending slightly inward due to the 1/16-inch gap).

With these two sides secured to the base, next mark and drill holes in the top cover (Fig. 11-26). We prefer clear acrylic to reveal the internal circuitry, but wood or opaque plastic also can be used. We have never had any problems with the acrylic top sheet failing, but if there is concern about its strength and durability, some supporting lips can always be bent onto the base sides (from Fig. 11-24) or some L-brackets added to support the top sheet. We don't do either of these because we have found the acrylic sheet held in place

by 12 screws to be plenty strong while providing a nice, open, and clean look at the base circuitry.

Also note that if a more brittle plastic is used, such as clear acrylic, the holes going into the sheet should be drilled and tapped to accept the bolts; otherwise, this more-brittle material will likely crack if screws are drilled into smaller holes.

Once the side holes in the top plate have been located, drilled, and tapped, set the top plate aside

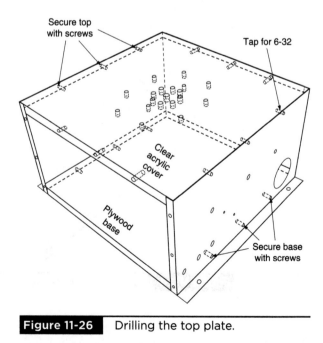

Figure 11-26   Drilling the top plate.

Figure 11-27   Base with sides and components (side plate removed).

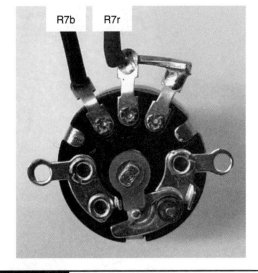

Figure 11-29   Wiring of dial R7.

so that the components can be secured inside the base either with screws or epoxy to hold them down (Fig. 11-27).

Wire the components with the front panel (Fig. 11-28).

Connect black wire R7b to the first pin of dial R7, and red wire R7r to the second pin (which is soldered to the third wire) as shown in Fig. 11-29.

5. Assembly: Chimney and Ladder

The chimney accomplishes three important things. First, it protects people from the arc, which is not life-threatening but can cause a painful burn if touched. Second, the chimney holds up the ladder (the rods)—these rods can be held up by other methods, but they are not self-standing and will need to be supported by something, and this chimney does a great job. Third, with a fan at the base, the chimney acts as a wind tunnel to help push the arc upward at

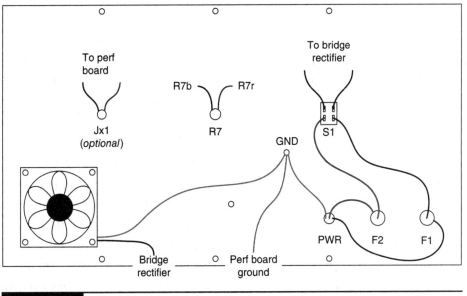

Figure 11-28   Front-panel wiring.

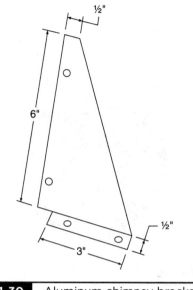

Figure 11-30   Aluminum chimney bracket.

a consistent rate, providing a better visual effect and preventing the arc from getting stuck on the ladder, where it will scorch a carbon deposit and become more likely to get stuck again in the same place.

To fabricate the chimney, first make four aluminum brackets (Fig. 11-30).

Next, cut a piece of 3/8-inch (or thicker) clear acrylic (Plexiglas or Lucite) into two strips, each about 8 inches wide and 48 inches long. Then cut two more strips of thinner 1/16-inch-thick acrylic (you may use the same thicker material, but this gets expensive). These four strips will make the chimney.

A tricky part to this assembly is getting the chimney sides and brackets all aligned flat and square as they rest on the top plate. The best way to do this is to stand the chimney sides up on the top plate as they will be mounted, and use tape along the edges to hold them together (such as masking tape or a strong painter's tape—something that will hold the sides together without leaving any sticky residue when removed). Start with taping two sides together (a thick sheet and a thin sheet), and run tape from the top down the edge (run some tape down the inside edge too if it feels a little loose). This piece should stand in place on its own while

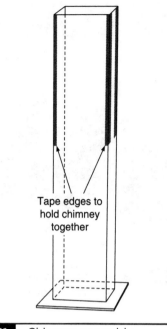

Figure 11-31   Chimney assembly.

you grab a third sheet and again run tape down the edge. Then tape the last sheet in place.

With the chimney sides held together and flat on the top plate, drill through the thinner 1/16-inch sheets and into the sides of the thicker 3/8-inch sheets. It is best to start at the bottom through the bracket holes. Tap for 6-32 screws, about 1 inch deep, spaced about 6 inches apart, and then secure the chimney together with screws. Be sure that the bottom two screw locations are aligned through the bracket holes (Fig. 11-32).

After all four brackets are secured in place at the bottom, work your way up the chimney, one side at a time, drilling and tapping about every 6 inches while making sure that the edge is squared. It is okay if the chimney top is not level on all four sides because a cap will cover this (see Fig. 11-36).

Once all sides of the chimney have been screwed together, center it on the top sheet, and then mark and drill the holes through the brackets into the top plate. These can be tapped for 6-32 screws to give a slightly better appearance, but the best method for stability is to drill a 3/16-inch hole and just run a bolt-washer-nut combination straight through

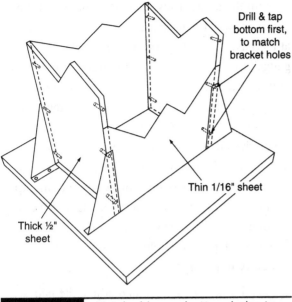

Figure 11-32 Match chimney bottom holes to bracket.

to firmly hold the bracket in place (no need to overtighten; just snug is fine).

But don't bolt the chimney to the top plate yet. Stand the chimney aside for now.

Instead, next secure the top plate in place with side screws, leaving off one side panel for access (Fig. 11-26).

Attach the two aluminum terminal blocks to the top plate with a bolt, washer and nut, attaching the HV transformer output leads WHV1 and WHV2 to the bottom and securing them in place (Fig. 11-33).

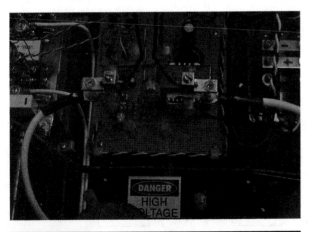

Figure 11-33 Ladder blocks secured to top plate.

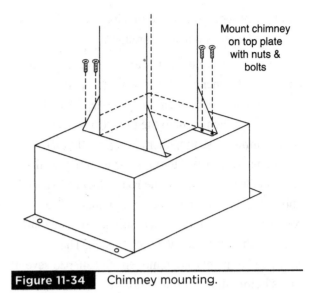

Figure 11-34 Chimney mounting.

Note that the ladder rods should not be put into the blocks at this stage.

Secure the chimney on the top plate, and then attach the remaining side panel (Fig. 11-34).

After the chimney has been secured to the top plate, remove the front screws of one of the thin 1/16-inch chimney panels to provide access for installing the ladder rods. Resecure the front brackets to hold the chimney steady and prevent it from tipping while work is being done.

The chimney rods will need to be bent inward at the base (Fig. 11-35). For now, they do not need to be secured, only bent to proper shape at the bottom

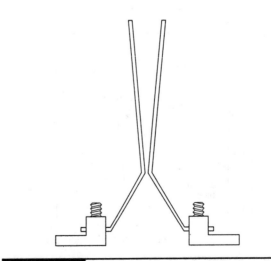

Figure 11-35 Ladder rods bottom shape.

and lightly secured into the blocks so that the rods next can be shaped and secured at the top.

Let the rods continue upward and slowly open to a gap about 3 inches wide at a height of about 5 ft tall. At the top, bend them in a 180-degree half-circle, and attach them to the thick (½ inch) sides of the chimney, seating them in terminal blocks that are themselves bolted through the chimney wall (Fig. 11-36). The tops of the rods should be spaced about 3 inches apart, so the bent diameter of the rods will vary with the exact chimney dimensions but also should be about 3 inches each. The rods may need to be placed, removed, and shaped several times before the final dimensions are correctly established. Drill holes in the chimney walls to secure the terminal blocks *after* the rods have been shaped to ensure that they are positioned correctly.

After the ladder rods are complete, fabricate a cap to cover the top of the chimney, which is just an aluminum sheet cut and bent to fit the chimney dimensions with a large hole (or holes) cut in the center to allow airflow (add some metal grating over the cut holes to improve their appearance, if required) and secured to the chimney by the topmost screws.

With the ladder rods secured at the top of the chimney, now the proper gap width can be set at the base for best spark initiation. Use a "gap gauge" (a piece of wood or anything ¼ inch wide will do), and slide the rods in the blocks until they reach this width (Fig. 11-37). Tighten down the rods to ensure a good electrical connection. When the Jacob's ladder is operated, this width may need to be fine-tuned for best operation: Too close and the arc may have difficulty moving up the ladder; too far and the arc may have difficulty forming. Usually the problem is initiation, so move this gap slightly closer if the unit has difficulty initiating the arc.

Test the Jacob's ladder, and adjust the gap of the ladder rods at the base until a consistent arc is generated. Then the chimney can be fully enclosed, and the Jacob's ladder is ready to use (Fig. 11-38).

The controls are simple and consist of the Power Switch S1, and the "On Time" dial R7, which determines the length of time that an arc stays lit before the next pulse; there is also a remote activation switch RJ1 (Fig. 11-39).

The optional remote activation switch may be installed to turn the ladder on from a distance (for

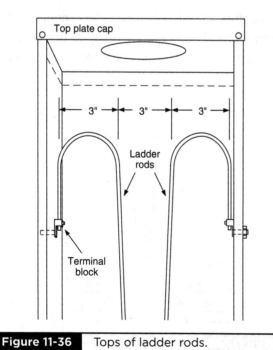

Figure 11-36 Tops of ladder rods.

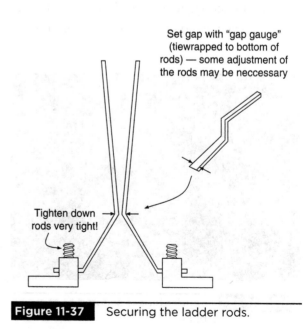

Figure 11-37 Securing the ladder rods.

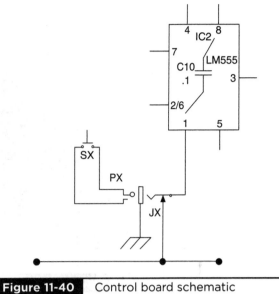

Figure 11-38 Chimney, brackets, and ladder assembled.

Figure 11-40 Control board schematic modification for remote switch.

use in demonstrations, interactive displays, haunted houses, etc.). This is a simple modification to the Control Board in Fig. 11-3 with a jack RJ1 on the front panel and two wires from the jack to interrupt pin 1 of IC2 (Fig. 11-40).

The jack itself is a closed-circuit design, where the circuit is closed until the plug is put into the jack to separate the connection (Fig. 11-41); notice how the fingers of the left pin are touching the plate of the right pin and

how inserting a plug into that jack will separate them—this is the type of closed-circuit mono jack you want for the remote switch. Then pressing the remote switch will close the circuit and activate the Jacob's ladder.

Figure 11-42 is a block diagram showing all the assemblies wired together.

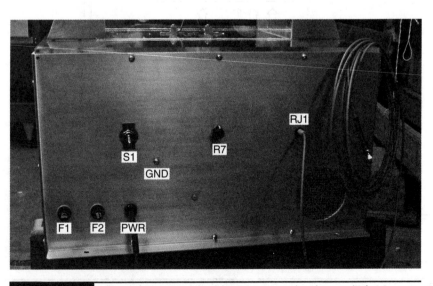

Figure 11-39 Base front panel with remote activation switch.

Figure 11-41   Closed-circuit jack.

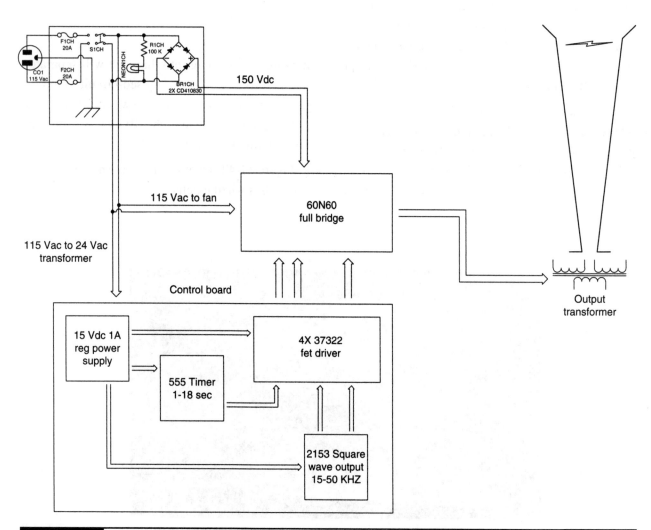

Figure 11-42   Block Diagram.

TABLE 11-1 Parts List			
Ref. No.	Quantity	Description	Part #
		Full Bridge	
R1–8	8	10-Ω 1-W film resistor (BR, BLK, BLK)	
C1, 2	2	1-µF 600-V polypropylene	
C3		6300-µF 400-Vdc vertical electrolytic	6300u/400V
D1–4	4	IN5819 19-V shotkey	
D5–12	8	IN4749 15-V zener	
D13–16	4	1.5KE30CA 1.5-kJ 30-V breakover diode	1.5KE/30CA
D17–20	4	MUR1660 dual fast-recovery diode	MUR1660CTG
D21–24	4	1.5KE400CA 1.5-kJ 400-V breakover diode	1.5KE/400CA
Q1–4	4	HGT1N60N60 600-V 60-amp high-powered fet	HGT1N60N60
TBLK1–4	4	Electrical Terminal Block (also called keystone terminals)	
T1, 2	2	10 turns of twisted-pair pri, 20 turns twisted pair secondary, core OP42212 (Ch. 10, Fig. 10-19)	JL6GATE
Fan	1	UF-80B11 115 Vac	FAN115/3
Heat sink	1	Aluminum C-beam 5-1/2" × 3" × 1-1/2" (Ch. 10 Fig. 10-4)	FBRHS
Chassis	1		
Screw			
		Solder lug	
		Controller Perf Board	
R1		10-Ω ¼-W film resistor (BR, BLK, BLK)	
R2		5-kΩ ¼-W film resistor (GRN, BLK, RED)	
R3		2.2-kΩ ¼-W film resistor (RED, RED, RED)	
R4		1-kΩ ¼-W film resistor (BRN. BLK, RED)	
R5, 8	2	22-kΩ ¼-W film resistor (RED, RED, OR)	
R6		500-kΩ vertical trimpot	
R7		500-kΩ 17-mm linear pot	
R9		10-kΩ ¼-W film resistor (BRN, BLK, OR)	
C1		1000-µF 35-V vertical electrolytic	
C2, 6, 7, 10, 11	5	0.1-µF 50-V ceramic disk	
C3, 8, 9	3	100-µF 25-V vertical electrolytic	
C4		0.001-µF 100-V polyester	
C5		0.0047-µF 100-V polyester	
C12, 14, 16, 18	4	10-µF 25-V vertical electrolytic	
C13, 15, 17, 19	4	0.1-µF 50-V ceramic disk	
C20, 21	2	10-µF 25-V vertical electrolytic	
D1		1N4937 1000-V 1-amp fast-recovery diode	
D2, 3	2	1N914 60-milliamp general-purpose signal diode	
D4, 5, 6, 7	4	1N5819 shotkey diode	1N5819
IC1		IR2153 high-side low-side driver 8-pin dip	IR2153
IC2		LM555 general-purpose timer 8-pin dip	

*Most parts should be available through electronics or hardware stores, but those more difficult to acquire are listed with a "Part #" and are available through www.amazing1.com if needed.

TABLE 11-1 Parts List (*Continued*)

Ref. No.	Quantity	Description	Part #
		Controller Perf Board	
IC3, 4, 5, 6	4	UCC37322P high-speed low-side mosfet driver with enable 8-pin dip	
BR1		KBPC806 bridge rectifier	
T1		LP-433 115-Vac in 24-Vac 1-amp out	
VR1		L7815 15-Vdc regulator	
SOCK8X	6	8-pin IC socket for IR2153, LM555, and UCC37322P	
		Miscellaneous Chassis Parts	
CO1		No. 14 three-strand line cord	
S1CH		DPDT	
F1, 2CH		20-amp fuse holder	
R1CH		100-kΩ ¼-W film resistor (BRN, BLK, YEL)	
NEON1CH		Neon indicator bulb	
BR1CH		2X CD410830 dual diode POW-R-Blok	
		17- × 17- × ½-inch plywood for base	
		17- × 17- × 3/8-inch (or larger) acrylic or plywood	
		17- × 17- × 1/16-inch plastic insulating sheet	
	2	4 × 7 cut in half diagonally to yield two brackets	
	2	17- × 17- × 1/16-inch aluminum for sides	
	2	18- × 18- × 1/16-inch aluminum for front and back	
	2	8- × 48- × 3/8-inch plexiglas or lucite	
	2	8- × 48- × 1/16- to 3/8-inch plexiglas or lucite	
	2	48- × 1/16-inch stainless rod	
	4	¼-inch terminal block electrode mounts	
		Output Transformer	
BASE		4- × 12- × ¼-inch lexan	
	4	4- × 2- × 1-inch square ferrite C core	CORE4
SEC INNER	2	2-3/8- × 1½-inch id tube	
SEC OUTER	2	2-3/8- × 3¼-inch od tube	
	2	2 inches × 50 ft mylar tape per coil	
	2	roughly 500 ft no. 24 magnet wire	
		two-part silicone elastomer	
		4' 20kv wire	
PRI BOX		6½- × 2¾- × 1/16-inch lexan	
		32 ft no. 16 litz	
		Assembled transformer as per Fig. 11-16, available as a ready-to-use module	JACK60TR
		Bridge Rectifier	
Dual diode	2	Half-bridge 800V 30A	BR800V30A-HB
Chassis	1	4-1/2- × 3-1/2-inch aluminum sheet	
Strip	2	2-1/8- x 1/2-inch aluminum strip	
Screw	4	1/2-inch 6-32	
Lock washer	4	6-32	
Screw	6	Matches half-bridge threads	
Terminal	4	14-gauge electrical terminal lug	
Terminal	1	Small electrical terminal lug	

CHAPTER 12

Free High-Voltage Experimental Energy Device

Overview

Obtain free high-voltage electrical energy from air currents using a corona-free probe and capacitor integrator. When installed properly (as shown in Fig. 12-1) in an active area, voltages of over 30 kV are possible. The current can charge a capacitor to a value producing energy of $W = \frac{1}{2}CVE^2$.

This is a demonstration of a means of obtaining free energy. But do not think that it will supplement your current energy demands: It will *not* do this. The concept is valid—free energy *can* be obtained from the atmosphere—but the actual amount obtained is small for the expense. Acres of land covered with these devices wired together would be needed to obtain perhaps 100 W of usable power. It would take literally centuries to recoup the expense of the investment. It is possible that future advances in materials science could reduce the cost to a point where this would become affordable, but as things stand today, the expense is too great for most to justify the return. However, the idea is sound—it works—and it could be expanded on in highly active areas to produce more energy or possibly modified to reduce the cost.

Air can become electrically charged as it moves over the earth's surface. These charges can be positive but are usually negative. The effect is similar to the charge picked up by a blimp airship that has been known to kill the first person grabbing a tether line if it has not been grounded properly. This same buildup of electrical energy also causes

corona discharges to sometimes occur on the masts of ships, the effect of which is known as *St. Elmo's fire*. These charges even have been known to cause swamp gas to produce an eerie glow that has been confused as UFOs. The charged air, as it moves, can be equated to a current source that charges a capacitive element by contact causing the voltage to rise to high values.

These plans show how you can obtain free energy from air currents in the form of high voltage that accumulates onto a capacitor. Under the right conditions, it is possible to obtain an appreciable amount of energy in joules. You should be able to obtain a healthy 20,000 to 30,000 V accumulated charge when air currents are electrified properly. You will note that at certain times, the effect is more active than at others. Early morning appears to be the best. Activity is noted by the flashing of a neon bulb connected in the system.

The charge is picked up by a metal probe that is nothing more than an insulated length of aluminum tubing. This tubing must be insulated and mounted to an insulated plastic arm that can swivel when attached to a vertical mounting pole. The ends of the tube contain metal spheroids to prevent electrical leakage. This scheme allows experimenting with a vertical or horizontal adjustment of the probe for maximizing effect. You should attempt to position the probe tube as high as possible.

You must use caution in a highly active thunderstorm area because lightning can be

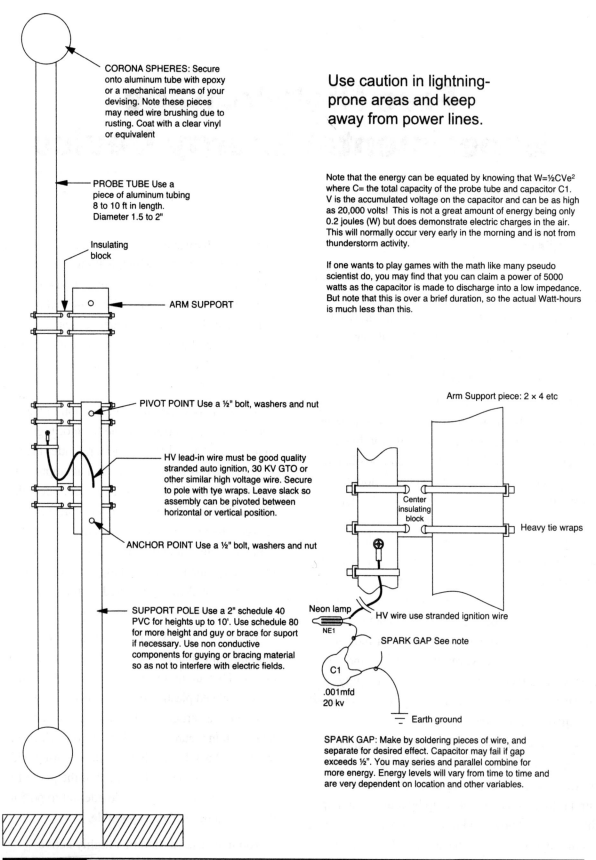

CORONA SPHERES: Secure onto aluminum tube with epoxy or a mechanical means of your devising. Note these pieces may need wire brushing due to rusting. Coat with a clear vinyl or equivalent

PROBE TUBE Use a piece of aluminum tubing 8 to 10 ft in length. Diameter 1.5 to 2"

Insulating block

ARM SUPPORT

PIVOT POINT Use a ½" bolt, washers and nut

HV lead-in wire must be good quality stranded auto ignition, 30 KV GTO or other similar high voltage wire. Secure to pole with tye wraps. Leave slack so assembly can be pivoted between horizontal or vertical position.

ANCHOR POINT Use a ½" bolt, washers and nut

SUPPORT POLE Use a 2" schedule 40 PVC for heights up to 10'. Use schedule 80 for more height and guy or brace for suport if necessary. Use non conductive components for guying or bracing material so as not to interfere with electric fields.

Use caution in lightning-prone areas and keep away from power lines.

Note that the energy can be equated by knowing that $W = \frac{1}{2}CVe^2$ where C= the total capacity of the probe tube and capacitor C1. V is the accumulated voltage on the capacitor and can be as high as 20,000 volts! This is not a great amount of energy being only 0.2 joules (W) but does demonstrate electric charges in the air. This will normally occur very early in the morning and is not from thunderstorm activity.

If one wants to play games with the math like many pseudo scientist do, you may find that you can claim a power of 5000 watts as the capacitor is made to discharge into a low impedance. But note that this is over a brief duration, so the actual Watt-hours is much less than this.

Arm Support piece: 2 × 4 etc

Center insulating block

Heavy tie wraps

Neon lamp

NE1

HV wire use stranded ignition wire

SPARK GAP See note

C1

.001mfd 20 kv

Earth ground

SPARK GAP: Make by soldering pieces of wire, and separate for desired effect. Capacitor may fail if gap exceeds ½". You may series and parallel combine for more energy. Energy levels will vary from time to time and are very dependent on location and other variables.

Figure 12-1   High-voltage experimental energy device.

hazardous, and we advise, under these conditions, that you remove the probe or securely ground it. Unlike a lightning rod, where the charge is made to leak off, the probe is designed to do the opposite, consequently increasing the lightning-strike hazard.

Basic construction is simple, and this project is easy to make, with most of the parts being available from your local hardware supply. Some parts are a little more special purpose and may be more difficult to source; if so, these can be obtained from www.amazing1.com as a kit for these specialized items (Part No. FREEHV1K).

Hazards

Minimal. The voltage is high, but the current is low and poses no shock danger unless used improperly (e.g., touching or standing nearby during an electrical storm) or scaled up to a much larger size.

Difficulty

Intermediate. Most of the difficulty is to be found in the size of this device because it stands about 12 ft high, but the actual assembly required is noncomplex basic construction, with a minimal amount of simple electrical work.

Tools

Basic construction and electrical tools.

Assembly

Prepare your materials. The support pole should be 2-inch Schedule 40 polyvinyl chloride (PVC) for heights up to 10 ft. Use schedule 80 for more height, and guy or brace (using nonconductive components so as not to interfere with electrical fields) for additional support if necessary. The required length of your support pole is half the length of your probe tube plus the diameter of a corona sphere plus at least 3 ft (1 ft for ground clearance and another 2 ft, or whatever is required, for anchoring the support pole into the ground). So a 10-foot probe tube with 4-inch spheres would require a support pole at least 8 ft 4 inches long. See Fig. 12-1 as an idea.

Drill ½-inch holes in the support pole for the pivot point and anchor point. Drill the pivot point about 3 inches back from the end of the support pole.

The arm support (Fig. 12-4) can be a piece of wood, such as a 2 × 3 or even a cut section from the support pole if you have extra length. As a rough guide, the arm support should be about one-fifth the length of the probe tube (so a 10-foot probe tube could use a 2-foot arm support). Drill one ½-inch hole in the center of the arm support and another at the end that lines up with the ½-inch hole in the support pole.

For the pivot point, use nylon lock nuts to prevent slippage. First, put the bolt through the support pole with a washer at each end, and tighten up the nut just until it is snug—no need to overtighten. Then put on another washer, the arm support, another washer, and the last nut (Fig. 12-2). This last nut you will want to tighten just to the point where the arm support can still rotate freely. The nylon threads will hold it in place so that the arm support neither binds up nor gets too loose.

In some situations, the electrical field may be shorted out by the vertical rod, and you may find that the probe tube is more effective in a horizontal configuration—this is where an optional arm brace comes into use (Fig. 12-3), a beam about 2 inches wide by 24 inches long that secures the probe tube horizontally. The actual length will depend on your build and how much space is between the pivot point and anchor points on the arm support (see Fig. 12-1).

The easiest way to figure out the required length for this arm brace is to simply rotate the arm support horizontally and then measure the distance between the anchor points in the arm support and

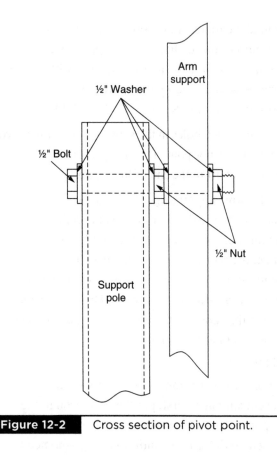

½" Washer

Arm
support

½" Bolt

½" Nut

Support
pole

Figure 12-2   Cross section of pivot point.

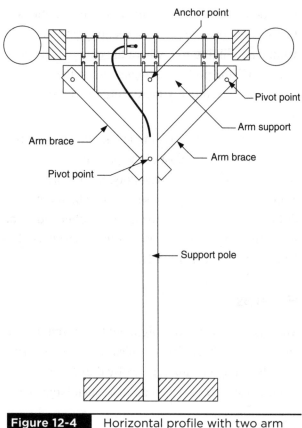

Anchor point

Pivot point

Arm support

Arm brace

Arm brace

Pivot point

Support pole

Figure 12-4   Horizontal profile with two arm
braces.

the support pole (Fig. 12-4). The brace should be
about the thickness of the middle bolt plus washers
between the support pole and the arm support
(see Fig. 12-2), which is likely ½ inch. Adjust the
width if you will be using two arm braces. Or you
can forgo the arm brace and just rotate the probe

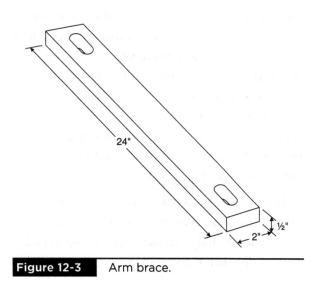

24"

½"

2"

Figure 12-3   Arm brace.

tube "upside down" such that the probe tube
hangs beneath the arm support and is held more
or less stable by its own weight. The drawback to
this method is that the probe tube might not bend
enough for the corona sphere to clear the support
pole, so in this case the corona sphere can be
removed and then replaced once the probe tube has
pivoted enough.

The best material to use for insulating blocks
is a 2-inch square bar or stick of PVC cut into
3-inch-long sections with four holes drilled
about ½ inch back from each corner through
which the tie wraps will be threaded (a drill press
works best here or clamps to secure the block,
allowing you to operate a drill with both hands for
straighter holes). If you can't find PVC blocks,
then wooden blocks also will work—but be aware
that eventually wood will rot and also will soak
up water in rainstorms, which may affect the
electrical characteristics of the probe tube.

The higher off the ground the device is located, the better. On top of your roof would be very good, if not for the danger involved during thunderstorms, so it is not recommended for this reason. If this device is to be secured into the ground, dig a hole with an auger or shovel for the support pole, at least 2 ft deep. For extra stability, the bottom may be filled with concrete because this is a nonconductive material. If you are unaccustomed to digging, be aware that 2 ft may not seem like much but requires a considerable amount of effort, especially in rocky soil, so pace yourself and do not cause excess strain.

After the support pole has been secured in place, run the bolts through the pivot and anchor points.

Attach the insulating blocks to the arm support, secured firmly in place by heavy tie wraps.

Crimp the end of your HV wire with an electrical connector, and secure this to the center of the aluminum probe tube with a sheet-metal screw. Hold the HV wire in place with a tie wrap to reduce strain on the crimp and screw. Wire to a neon lamp NE1 for visual confirmation of electric charge and then to a capacitor (or capacitor bank) with a spark gap, as shown in Fig. 12-1.

TABLE 12-1 Parts List*		PART #
1	Support Pole: 2-inch-dia Sched 40 PVC or Sched 80 if over 10 ft high	
1	Probe Tube: 1.5- to 2-inch-diameter, 8- to 10-foot length of aluminum tubing	
2	Corona Sphere: 4 -inch-diameter (or any round metal object)	SHP4STEEL
3	Insulating Block: fabricate from PVC or wood	FREEHVBLKFAB1
15	Heavy tie wraps, ¼ inch wide by 1/16 inch thick	
1	1-inch × 3-inch × 2-foot Arm Support (wood)	
1–2	½- × 2- × 24-inch Arm Brace (PVC or wood)	
2–3	½- × 5-inch bolt	
6	½-inch washer	
5–6	½-inch nut (nylon "locking")	
1	0.001-µF 20-kV capacitor (0.001µF/20-kV)	.001u/20KV
1	Neon bulb NE1	NE1
1	HV wire	WIRE20KV

Note that the SHP4STEEL pieces may need wire brushing due to rusting. Coat with a clear vinyl or equivalent.

*Most parts should be available through electronics and/or hardware stores, although some may be more difficult to acquire, and these are listed with a "Part #" and are available through www.amazing1.com if they prove difficult to find elsewhere.

CHAPTER 13

HHO Power Conditioner

Overview

These instructions show how to build a power conditioner (Fig. 13-1) that generates the electrical output that splits liquid H_2O into gaseous H_2 and O_2. The electrical output goes into a reactor cell (Chap. 14) in which the electrochemical reaction known as *electrolysis* physically occurs.

Note that output is 12 to 14 V at 30 amps, with current being controlled by *current chopping*, otherwise known as *duty cycle control* (see Chap. 22 for more information).

Hazards

The only hazards associated with building this reactor cell are the normal construction precautions involved with soldering and drilling/cutting steel and plastic. This device itself does not directly produce HHO gas, but it does provide the electricity used for electrolysis, and the resulting HHO gas generated in a reactor cell *is* very dangerous, increasing with the amount of gas produced. Great caution should be taken when working with HHO gas, such as using multiple flash-stop chambers (bubblers), avoiding open flames or anything that may produce an electrical or static discharge, and only producing a limited amount of the gas at any one time. **Eye protection should be worn when making, testing, and operating this device.**

Difficulty

Intermediate. Basic fabrication skills and intermediate electrical soldering skills are required.

Tools

Drill, metal and wood saws, 6-32 screw tap, pliers, soldering gear, fabrication equipment.

Hydrogen Gas Production

The next seven chapters involve hydrogen gas production or, more precisely, HHO gas production, which is much more dangerous than hydrogen gas because the oxidizer (in this case, pure oxygen) is already mixed into the hydrogen gas at the ideal ratio for rapid combustion.

This chapter shows how to make an *HHO power conditioner*, which is the electronics that converts water (H_2O) into an explosive gas (HHO).

Figure 13-1   Assembled HHO power conditioner.

243

This conversion process is called *electrolysis* (from the Greek *lysis*, which means "to loosen," so *electrolysis* simply means to use electricity to "loosen up" hydrogen and oxygen atoms from water molecules).

Chapter 14 shows how to make an HHO reactor cell, which is the physical reactor core where the electrolysis takes place. Chapter 15 is a variation on the HHO reactor cell that produces hydrogen gas faster and more efficiently, but at a lower maximum gas pressure. Chapter 16 shows how to make an igniter, which produces a controllable high-voltage spark that is used to set off the HHO gas.

These three components—the power conditioner, which creates the proper electrical conditions where electrolysis occurs; the reactor, where electrolysis physically takes place; and the igniter, which detonates the HHO gas—are necessary for HHO gas creation, and then a controlled combustion.

The next four chapters show you how to make some devices for having fun with HHO. Chapter 17 shows how to make an HHO bomb, which is simply a bag filled with HHO gas that is detonated from a safe distance. This can be used as a straightforward boom and flash for percussional pyrotechnic effects, or it also can be used to launch items into the air for gravity-defying amusement (our half-gallon bags have launched 5-gallon buckets well over 100 ft straight up). Chapter 18 shows how to make an HHO pistol, which is, of course, a handheld device that launches smaller projectiles. Chapter 19 shows how to make an HHO howitzer, a sizable cannon capable of firing a 50-mm projectile at high velocities and to distances of well over 1000 ft. This is a dangerous item, but it is also great fun!

Notes on HHO Gas

Now at last the ultimate all-action science project using only "harmless" water for the energy source!

WARNING   The projects in the next seven chapters can be dangerous and are intended to be supervised by qualified school, university, or adult personnel. The blast effects can cause damage and injury.

Unfortunately, we have no control over the amounts of the gases or the types of blast containers used, nor can we enforce our safety recommendations. We therefore can only warn experimenters to use small amounts in plastic baggies and to treat these devices as real weapons. Experimenter must also be aware of the fact that this is the most efficient form of released chemical energy per given weight of any conventional explosive.

But do not get your neighbors or the proverbial worrywarts into a dither. The project only demonstrates the potential *chemical* energy from water. This energy is easily released from ordinary water. Yes, only ordinary water is used to produce these extremely attention-getting results! Fortunately, a real thermonuclear bomb derives energy from the strong *nuclear* forces. This action requires intense energy usually such as that of a fission bomb. This is necessary for generating the temperatures required to overcome the intense Coulomb forces encountered in trying to force the hydrogen nuclei to fuse into helium. This is where the strong forces come into play releasing oodles of giga-gigajoules. Technically, all elements up to but not including iron can fuse under certain conditions such as mega-temperatures released by such giga-gigajoules of energy—*kind of scary when you think about it! Note that it takes the energy of a supernova to produce the heavier elements.*

Even our engineering department might find the fusion process difficult without a small fission bomb! These simple fission devices take little relativistic energy to destabilize the nucleus but do take time in processing to a high enough grade.

Our research programs are now attempting to temporarily neutralize the Coulomb forces within the tritium nucleus, thus allowing less thermal energy to cause the fusion process. These extremely fast high-energy electric pulses must produce a very high-voltage standing wave to appear. "That's all folks" we can elaborate on at this time.

Hydrogen and oxygen are produced by simple electrolysis that used to be taught in elementary school. The methods shown were very inefficient and produced only enough material to show an orange flame when ignited. Thanks to pioneers such as Stan Meyers and others, groundwork was laid to produce hydrogen faster and more efficiently. Larger amounts produced over a shorter period of time allow more ease in performing experimental research and other useful functions.

Using hydrogen to fuel automobiles is the objective of major research currently being performed by many companies as well as individuals. Hydrogen, when burned, combines with oxygen to produce water and heat. The problem is that the chemical energy of burning cannot exceed the amount of electrical energy that is required to convert the water into the individual gases. Nuclear-produced electricity, in turn, used for electrolysis could provide a viable method of eliminating the nasty by-products obtained from fossil fuels.

Assembly

Figure 13-2 provides the circuit schematic. First, cut, drill, and bend aluminum sheet to make the chassis (Fig. 13-3). The front-panel holes don't need to be placed exactly but should be cut to

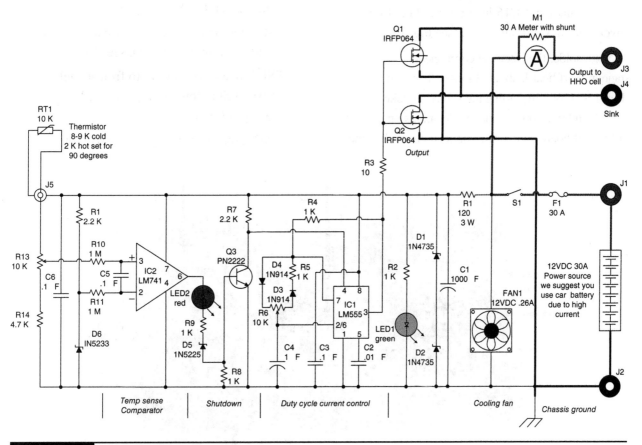

Figure 13-2   Circuit schematic.

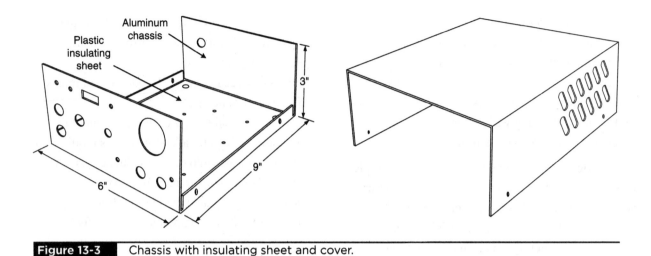

Figure 13-3 Chassis with insulating sheet and cover.

suitable locations for their respective controls (see Fig. 13-14). Cut an insulating sheet from plastic to fit inside the chassis, and drill holes—again, these don't need to be exact but are needed only to hold the components and boards in place (see Fig. 13-13). Then cut and bend 1/16-inch plastic sheet for the cover, adding ventilation slots (see Fig. 13-3).

Next, solder the electrical components onto a standard perf board, about 2½ × 4½ inches in size (Fig. 13-4). Before starting the soldering, take a marker and trace out component locations on one side of the perf board and wiring connections on the other.

Perf Board Wiring Connections

PL1: red wire going to front-panel Power Lamp (green LED; see Fig. 13-11)

PL2: green wire going to front-panel Power Lamp (green LED; see Fig. 13-11)

TSL1: red wire going to front-panel Thermal Shutdown Lamp (red LED; see Fig. 13-11)

TSL2: black wire going to front-panel Thermal Shutdown Lamp (red LED; see Fig. 13-11)

GND: green wire going to chassis ground

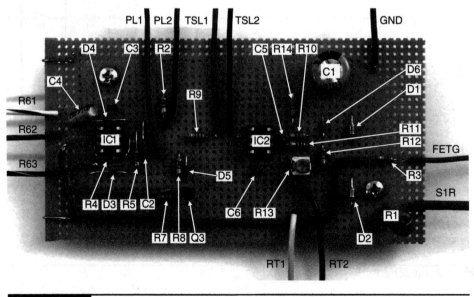

Figure 13-4 Perf board.

FETG: red wire going to field-effect transistor (FET) board (see Fig. 13-7)

S1R: red wire (16 AWG) going to front-panel switch S1 (see Fig. 13-14)

RT1: yellow wire going to rear-panel thermistor jack J5

RT2: blue wire going to rear-panel thermistor jack J5

R61: white wire with black stripe going to Current Adjust Dial R6 (see Fig. 13-10)

R62: red wire going to Current Adjust Dial R6 (see Fig. 13-10)

R63: black wire with white stripe going to Current Adjust Dial R6 (see Fig. 13-10)

The component soldering is shown in Fig. 13-5.

Figure 13-6 shows the aluminum heat sink with drilled holes, cut from a section of C-beam. The holes going through the body are simply straight-through 5/32-inch-diameter holes. The two holes at the corners are tapped to 6-32 threads for securing the fan to the heat sink. If the heat sink is narrower than the fan is wide, an aluminum shim or two (just a piece of 1½- × 3-inch aluminum sheet) can be placed between the heat sink and the chassis as spacers to ensure that the heat sink

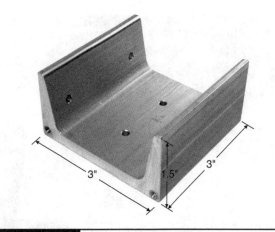

Figure 13-6   Aluminum heat sink.

has flush contact with the chassis for good heat dissipation. The heat sink is attached to the chassis with nuts and bolts going through the drilled holes (tighten snugly, but there is no need to overtighten).

Then fabricate the field-effect transistor (FET) board as shown in Figs. 13-7 and 13-8.

FET Board Wiring Connections

FETG: red wire going to perf board (see Fig. 13-4)

FETS: heavy white wire going to front-panel ground (see Fig. 13-12)

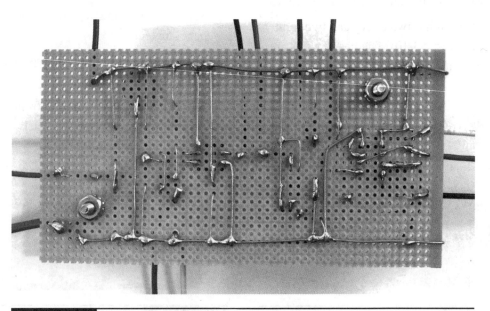

Figure 13-5   Perf board wiring.

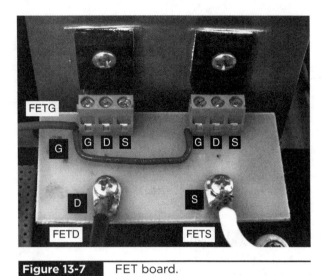

Figure 13-7 FET board.

FETD: heavy black wire going to front-panel negative reactor output jack (see Fig. 13-12)

Attach it to the heat sink (Fig. 13-9). A second heat-sink fin can be cut and bent from 1/16-inch aluminum sheeting for further heat dissipation, as shown. Also note the bent piece of aluminum that looks like a ramp secured to the top of the fan, which concentrates the airflow onto and across the heat sink to significantly improve the cooling rate. This is simply a sheet of 1/16-inch aluminum bent so that it can attach to the fan (with 6-32 tapped holes) and then sloped downward like a ramp to direct airflow (the profile can be seen in Fig. 13-13).

The current shunt has two heavy and two light wire connectors (Fig. 13-10). The heavy wires are connected one to the main power switch (MPS)

Figure 13-9 FET board attached to heat sink.

and one to a reactor output jack. The light wires are connected to the meter. The thin wire connected to the *negative* side of the meter must be on the same side of the current shunt as the heavy wire connected to the *positive* reactor output jack. The positioning of the shunt on the chassis can be seen in Figs. 13-11 and 13-13.

Current Shunt Wiring Connections

MPS: heavy red wire going to Main Power Switch on front panel (see Fig. 13-11)

M1: red wire going to *positive* side of front-panel meter

M2: red wire going to *negative* side of front-panel meter

RO: heavy red wire going to front-panel *positive* (red) reactor output jack

Figure 13-8 FET board component wiring.

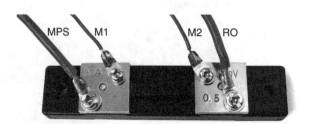

Figure 13-10 Current shunt.

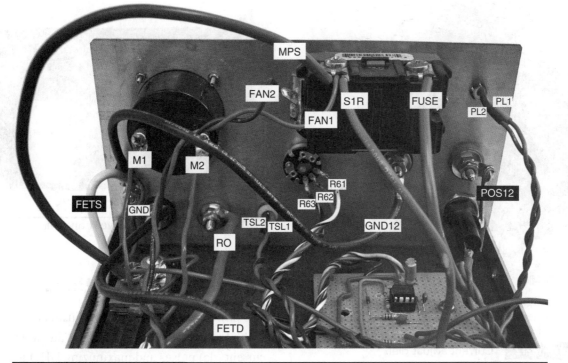

Figure 13-11 Front-panel wiring.

Next, connect and wire the front-panel components (Fig. 13-11).

Front-Panel Wiring Connections

MPS: heavy red wire going to current shunt (see Fig. 13-10)

S1R: red wire (16 AWG) going to perf board (see Fig. 13-4)

FAN1: red wire going to cooling fan

FAN2: black wire going to cooling fan

FUSE: heavy red wire going to front-panel fuse (center connector)

FETS: heavy white wire coming from front-panel chassis ground and going to FET board (see Fig. 13-7)

FETD: heavy black wire coming from front-panel negative output jack J4 and going to FET board (see Fig. 13-7)

M1: red wire on *positive* side of front-panel meter going to current shunt (see Fig. 13-10)

M2: red wire on *negative* side of front-panel meter going to current shunt (see Fig. 13-10)

GND: green wire going to perf board (see Fig. 13-4)

RO: heavy red wire going from current shunt to front-panel positive output J3 (see Fig. 13-10)

TSL1: red wire going to perf board (see Fig. 13-4)

TSL2: black wire going to perf board (see Fig. 13-4)

R61: white wire with black stripe going to Current Adjust Dial R6 (see Fig. 13-12)

R62: red wire going to Current Adjust Dial R6 (see Fig. 13-12)

R63: black wire with white stripe going to Current Adjust Dial R6 (see Fig. 13-12)

PL1: red wire going to perf board (see Fig. 13-4)

PL2: green wire going to perf board (see Fig. 13-4)

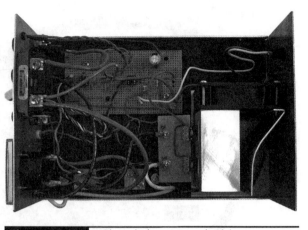

| Figure 13-13 | Chassis layout and wiring. |

Figure 13-12   Current adjust dial.

GND12: black wire going from front-panel
12-V negative input jack J2 to chassis ground
(also on front panel)

POS12: heavy buss wire going from front-panel
12-V positive input jack J1 to front-panel fuse
(side connector)

The Current Adjust Dial uses three of the five
connectors (Fig. 13-12).

Current Adjust Dial Wiring Connections

R61: white wire with black stripe going to perf
board (see Fig. 13-4)

R62: red wire going to perf board
(see Fig. 13-4)

R63: black wire with white stripe going to perf
board (see Fig. 13-4)

The overall layout and wiring connections are
shown in Fig. 13-13. Also note the profile of the bent
aluminum ramp that helps to direct the fan airflow
over the heat sink to improve heat dissipation.

And the front- and rear-panel controls and
connections are shown in Figs. 13-14 and 13-15.

Operating Instructions

The power conditioner allows adjustment of the
width to dwell time for controlling the direct-
current (dc) pulses and maximizing HHO gas
production. This feature matches the input power
to your reactor cell, eliminating overheating
and maintaining good production efficiency. A
temperature sensor prevents the system from
overheating. Large volumes of HHO gas now can
be produced. Simply attach the *power conditioner*
to the *reactor cell* to *bubblers* to *your project*. Turn
on the power, and wait for the meter to read 2 to
5 pounds of pressure; then admit the HHO gas to
your project. The Current Adjust can be adjusted
for the desired amps and rate of production.

**Our lab system can generate 50 lb/in² of
pressure, but this is a very dangerous amount
that can explode, causing damage and injury.**
We entirely discourage storage of this explosive
HHO gas. (If you do decide to try storing this
explosive gas, we recommend excavating a hole
in the ground deep enough in which to place
the storage tank, preferably ringed with a height
of sand bags such that any flying shrapnel is
prevented from sideways motion. There should be
two safety shutoff valves along with two water-
bubbler isolation chambers to arrest any flame
from possibly getting back to the storage tank and
causing a real nasty explosion. You might use

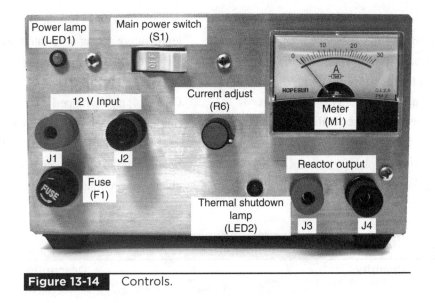

Figure 13-14 Controls.

turpentine in one of the bubblers. Setup and operation should prevent any necessity of going near the storage tank area. All precautions to eliminate the slightest static charge must be taken. Use copper tubing buried underground, and remotely locate the filling and dispensing areas—*and may God be with you!* We emphasize again: Do *not* store HHO gas. But if you are stubborn-headed and insist, at the very least, follow these safety precautions for the explosion that will eventually occur.

Figure 13-15 Rear-panel with temperature sensor attached.

Setup

1. Obtain a freshly charged automotive battery or a 12- to 14-V 30-amp dc power supply, and connect it to a power source. *Do not use a battery charger, which are 14–20 V output.*

2. Note the heavy-duty banana jacks and mating plugs. It will be necessary to connect and solder these plugs to at least 16-gauge wire leads. If lengths are to exceed 10 ft, you will need 14-gauge wire.

3. Attach the temp sensor to the midsection of the reactor cell using several wraps of electrical tape. Note that these wire leads are fragile.

4. Verify that the reactor cell is operating properly and that all is well ventilated. Connect the leads to the terminals on the base of the reactor.

5. Rotate the Current Adjust knob (R6) fully counterclockwise. Attach the 12-Vdc input (J1 and J2). Note whether the meter is reading. Rotate the Current Adjust knob to 15 amps as read on meter. Note the bubbles (they look like a fine white mist) of HHO gas forming in reactor cell. Let it run, and monitor the pressure gauge on the reactor cell. Do not allow the pressure to go over 25 lb/in^2 at this stage.

6. You may quickly generate the HHO gas by manually controlling the system for short-term use. This will require manually checking the reactor cell temperature, generated pressure, and current drive. This method is for immediate use of the generated gas for demonstration and so on.

7. To prevent accidental explosion by overheating, attach the temperature sensor to the reactor cell chamber. This is done by plugging the temperature sensor wire into the back of the HHOPC10 and then securing the temperature sensor head to the outside of the Reactor Core (see Chap. 14; tape wrapped around the reactor works fine). Once the temperature reaches 90 to 100°F, a signal is sent to the power conditioner to shut down until the reactor cell cools. This reduces the chance of spontaneous detonation by overheating.

TABLE 13-1 Parts List*			
Ref. No.	Quantity	Description	Part #
R1		120-Ω 3-W MOX noninductive (BR, RED, BR)	
R2, 4, 5, 8, 9	5	1-kΩ ¼-W film resistor (BR, BLK, RED)	
R3		10-Ω ¼-W film resistor (BR, BLK, BLK)	
R6		10-kΩ 17-mm vertical mount	
R7, 12	2	2.2-kΩ ¼-W film resistor (RED, RED, RED)	
R10, 11	2	1-MΩ ¼-W film resistor (BR, BLK, GRN)	
R13		10-kΩ trimpot	
R14		4.7-kΩ ¼-W film resistor (YEL, PUR, RED)	
RT1		10-kW thermistor (Mouser 334-NTC103-RC)	THER10k
C1		1000-µF 25-V vertical electrolytic	
C2		0.01-µF 25-V disk capacitor	
C3, 5, 6	3	0.1-µF 50-V disk capacitor	
C4		1-µF 25-V vertical electrolytic	
D1, 2		1N4735 6.2-V zener	
D3, 4		1N914 high-speed diode	
D5		1N5225 3-V zener	
D6		1N5233 6-V zener	
Q1, 2		IRFP064 *N*-type fet	IRFP064
Q3		PN2222 *PNP* transistor	
IC1		LM555 8-pin timer	
IC2		LM741 8-pin dip	
LED1		Green indicator led	
LED2		Red indicator led	
M1		30-amp meter and shunt	MET30AL
SOCK8X	2	8-pin connector for LM555 and LM741	
SOCK247		3-pin connector Mouser no. 158-P02ELK508V3-E for IRFP064	SOCK247
	2		
J1, 3		Red banana jack	

*Most parts should be available through electronics or hardware stores, but those more difficult to acquire are listed with a "Part #" and are available through www.amazing1.com if needed.

Ref. No.	Quantity	Description	Part #
TABLE 13-1 Parts List (*Continued*)			
J2, 4		Black banana jack	
J5		2.5-mm dc jack	
P5		2.5-mm dc plug	
FAN1		12-vdc 0.26-amp PN:3110KL-04W-B47	FAN1153
S1		Min 20-amp wall switch	
F1		30-amp fuse and holder	
		Extruded Aluminum Channel Heat Sink	
		Heat-sink plate	
		Heat-sink spacer	
		Heat-sink duct	
		Therma pads for IRFP064	
		Chassis	
		Insulator	
		Cover	
		Feet	
		Screws	
		Nuts	
		Optional fully assembled HHO Power Conditioner	HHOPC10

CHAPTER 14

HHO Reactor Cell

Overview

These instructions show you how to build a Reactor Cell that can produce copious amounts simultaneously of H_2 and O_2 gas from ordinary water (Fig. 14-1). This Reactor Cell is where the electrochemical reaction of *electrolysis* physically occurs. A power supply (such as the HHOPC10 Power Conditioner of Chap. 13) is required to provide the electricity for splitting liquid H_2O into gaseous H_2 and O_2.

Hazards

There are no hazards associated with the fabrication of this reactor cell, other than the normal construction precautions involved with drilling and cutting steel and plastic tubing. However, once the reactor cell is assembled and running, the HHO gas produced is potentially very dangerous, and the danger increases with the amount of gas.

Difficulty

Intermediate. Basic fabrication skills are required, such as cutting, drilling, tapping holes to match bolt threads, basic soldering, and general construction techniques.

Tools

Drills, metal and wood saws, 6-32 screw tap, pliers, soldering gear, PVC cement, 3M super-weather strip adhesive (yellow in color), glue, Teflon tape, Mylar tape, fabrication equipment.

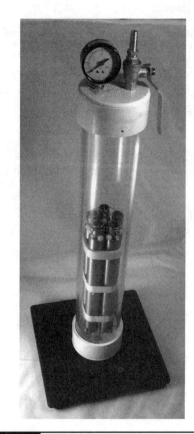

Figure 14-1   Assembled reactor cell.

Legal Notes

Given the climate of the world today, it would be wise to contact your local fire department, police department, and neighbors to let them know that you will be doing some experimenting with hydrogen gas and that they may hear some bangs as a result and so not to be alarmed. There also may be local ordinances against the use of HHO gas in an explosive manner.

Be sure to clear the area of any flammable objects (e.g., leaves, dead grass, paper, etc.), and have a hose and/or water buckets on hand just in case.

WARNING The author, publisher, and any and all individuals associated with these plans assume no responsibility for their use or misuse. If you build this device, then you are agreeing to accept full responsibility for any and all outcomes of its use.

WARNING Dangerous product when assembled, produces explosive gases.

We cannot stress enough the danger of accidental ignition of the HHO gas causing an explosion. All precautions involving static electricity and providing unimpeded outside ventilation are strongly recommended.

DANGER Unit easily generates 50 lb/in² in a volume of 50 in³ in several minutes.

Even though this unit as shown is built to withstand 80 lb/in² of pressure, for safety reasons, do not allow pressure to exceed several pounds per square inch. HHO by itself is an explosive gas, and putting it under pressure only compounds the danger.

Do not store this explosive HHO gas, even if you are familiar with handling and containment.

If you are considering storage, it would be a good idea to download the referenced Patent No. 308,276.

We suggest that you use our high-current HHO power conditioner (the HHOPC10 in Chap. 13) with thermal feedback control for maximum output. The HHO gas can be used for the HHO howitzer, mortar, bomb, and pistol and as a fuel supplement for gas engines, torches, rockets, and so on.

These instructions show you how to build a reactor that can produce copious amounts simultaneously of H_2 and O_2 gas from ordinary water. This combination, when ignited, produces a powerful explosion that can be used to power a wide variety of devices. One advantage of this energetic reaction is that the by-product is pure water. This eliminates the pollutants found in other chemical explosions, such as from coal and oil. Another advantage is that the energy yield per molar weight is the highest possible by conventional chemical reaction. This is why hydrogen and oxygen are used as a rocket propellant.

As with all forms of stored energy, it takes more energy to produce the gases from the water than the energy released on recombination back into water. (To overcome this disadvantage would be the Holy Grail of the energy problem.) One viable approach to producing cheap hydrogen gas is to use nuclear power to generate the electricity for electrolysis, supplying virtually unlimited amounts of hydrogen gas. If a very efficient reactor is developed, this energy could be used for pollutant-free fuel for all our transportation needs, helping to save the planet from global warming, acid rain, and other pollution-related problems.

When generating and igniting HHO gas, some form of flashback protection is required. Standard flashback protectors will not work because the flame front of the ignited HHO gas is 1000 times faster than that of gasoline. A simple solution is two or more *bubblers* in series with your experiment and gas supply, placed as close as possible to the experiment (which minimizes the amount of gas that might ignite in the tube).

A bubbler is shown on Figure 14-36 and can be made in different ways but must isolate the gas flow.

Setup (Perform Outside)

1. Fill the system with distilled water about 6 inches from the top. Add a pinch of salt to the water. You can fill the assembly with presalted distilled water by backfilling through the ball valve.

2. Condition the cell by adjusting the input current from the Power Conditioner to 1 amp, and allow it to run for 30 minutes, then shut it down

for 30 minutes. Change the water. If you note a brown sludge, it must be cleaned using a brush. Be careful not to touch the cells with your hand because they should be kept free of any oils that would impede the electrolysis.

3. Repeat above at 5 amps for 20 minutes, and shut down for 30 minutes.

4. Repeat above at 10 amps for 15 minutes, and shut down for 30 minutes.

5. Repeat above at 15 amps for 10 minutes, and shut down for 30 minutes.

6. Repeat above at 20 amps for 5 minutes, and shut down for 30 minutes.

CAUTION Do not allow the pressure to exceed 5 lb/in^2 as read on meter. Vent off excess pressure with the ball valve.

DANGER Escaping gas is prone to explosion by static electricity. Preferably perform on a wet or humid day. Ground yourself with a grounding wire (generally available at computer and electronics stores, both locally and online) to prevent static buildup and accidental discharge.

CAUTION Do not to allow the cell to get hot.

Once conditioned, the sludge will disappear, and the HHO gas production will increase.

Assembly

Start the assembly by cutting, drilling, and bending four aluminum base brackets (Figs. 14-2 through 14-4).

Drill 1/8-inch holes into the PVC base to prevent splitting when the screws are inserted (Fig. 14-5).

Center the PVC bottom base on a plywood stand, and secure it in place with screws (do not yet fully tighten them because they will be removed in a later step for modification to the PVC bottom base). Plywood should be ¾ inch thick (Fig. 14-6).

Figure 14-2   BB1 base brackets: four pieces of No. 22 galvanized steel.

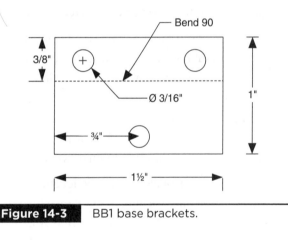

Figure 14-3   BB1 base brackets.

Figure 14-4   BB1 base brackets in position.

Figure 14-5 Bottom base.

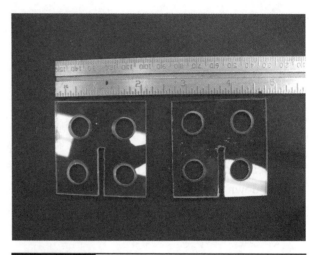

Figure 14-7 Cell stand brackets: 1/8-inch Plexiglass or polycarbonate.

Fabricate two cell stand brackets (Fig. 14-7). The holes are to allow the motion of water, and so they do not need to be perfectly centered.

Cut five pieces of ¾-inch outside diameter (OD) stainless steel tubing 11 inches long. Repeat and cut another five pieces with 1-inch OD stainless steel tubing. Clean the tubing with a Scotch Brite or Brilla pad inside and out (Fig. 14-8). Tubes must be very clean before operating. Use acetone to clean off oily fingerprints before final assembly.

With all 10 tubes made, mark hole locations (Fig. 14-9). Note the one hole in the 1-inch tube is ¼ inch from the end. The two holes in the ¾-inch tube are ¼ inch from the end with the second hole 5/8 inch from the first.

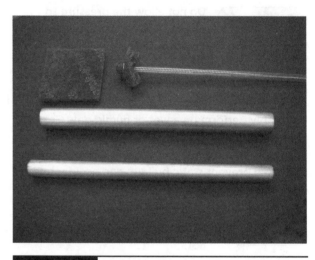

Figure 14-8 Initial prep of reactor tubes.

Figure 14-6 Wooden base.

Figure 14-9 Mark for drilling 3/32-inch holes.

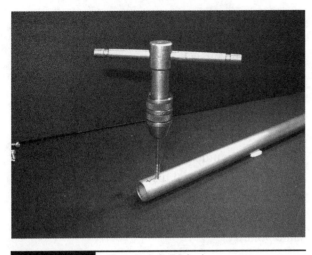

Figure 14-10 Tapping 6-32 holes.

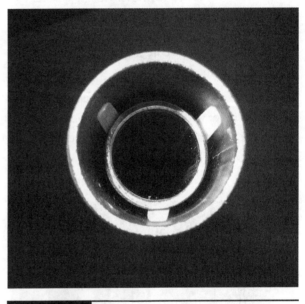

Figure 14-12 Tie wrap spacer placement.

Nylon screw
Tie wraps

Tap all holes using a 6-32 tap (Fig. 14-10).

Put 6-32 × 3/8-inch nylon screws in the second hole of the ¾-inch tubes (Fig. 14-11).

Place three tie wraps along the ¾-inch tubes at 120-degree intervals, and slide them into the 1-inch tubes (Fig. 14-12). Tubes must be perfectly coaxial and must not touch in any way. Spacing the tie wrap ends evenly at 120 degrees will prevent the tubes from touching (Fig. 14-13).

The outer 1-inch tube will stop at the nylon screw (Fig. 14-14).

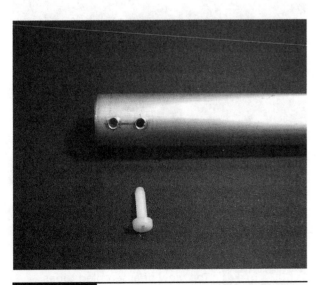

Figure 14-11 Nylon screw insertion.

Figure 14-13 Bottom view of tube spacing alignment.

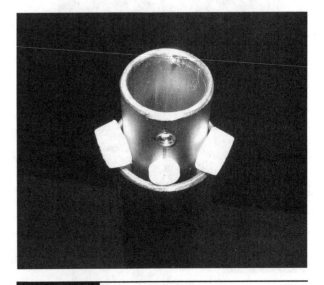

Figure 14-14 Assembled tubes (five required).

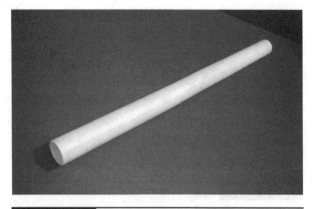

Figure 14-15 Center plastic tube 12 × ¾ inch inside diameter (ID).

The next step of the core assembly starts with the center plastic tube (Fig. 14-15).

Stand the metal tubes on a flat surface around the plastic tube, with the screw holes facing out. This is the orientation for assembling the tube cluster together (Fig. 14-16).

With the stainless dual tubes arranged around the plastic center tube, use elastics (rubber bands) to hold all the tubes together so that the outer steel tubes can be rotated around until everything aligns snugly in place before the tube assembly is secured together with tape (Fig. 14-17). You will have to pivot the tubes such that the tie wraps of each stay at equidistant separation (120 degrees) to keep the inner metal tube from touching the outer metal tube (it does not matter if the outer tubes touch each other because they are all the same polarity).

Once the tubes are positioned properly, vinyl tape will hold the assembly more firmly in place (Fig. 14-18).

Wire the tubes together using No. 18 solid stainless steel buss wire, wrapping once around each stainless steel screw keeping the wire fairly snug (but not overly tight), and tighten down each screw except one, which will be the connection point for the power leads. Twist the end of the stainless wire together with pliers, again keeping the wire snug but not overly tight (Fig. 14-19). The tie wraps may reposition themselves as the stainless wire is tightened, so do your best to keep them

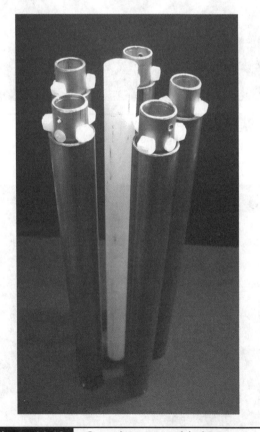

Figure 14-16 Complete assembly in proper position.

Figure 14-17 Top view of assembled tubes.

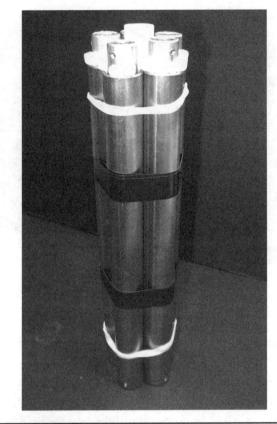

Figure 14-18 Tube assembly showing method of attaching together.

relatively equidistant. Repeat for the bottom section. Note that at this point in the construction, the tube assembly now can be referred to as the *core*.

Tape the tubes together as shown in Fig. 14-20 using 3- to 4-inch Mylar tape (use three or four

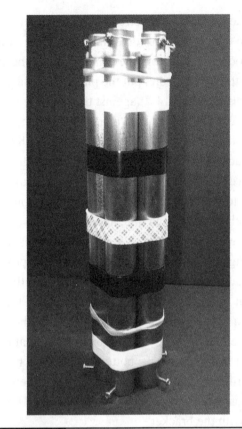

Figure 14-20 Final taping using ¾-inch Mylar tape.

layers), at which point the rubber bands can be removed.

Make a positive electrode from a piece of ½-inch-wide stainless steel (Fig. 14-21). Bend it as shown, and this bend will be on the top of the core.

Figure 14-19 Core wiring.

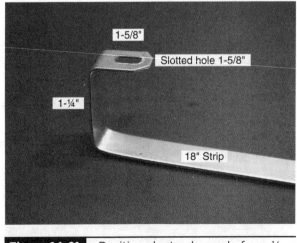

Figure 14-21 Positive electrode made from ½- × 1/16- × 18-inch length stainless steel.

Slide the positive electrode into the center plastic tube at the top side of the core. Where the electrode emerges from the center plastic tube at the bottom of the core, put the plastic cell stand brackets (see Fig. 14-7) against the bottom of the core for reference, and mark on the positive electrode where it should be bent to lay flush with the bracket. Once the bend line is marked, the positive electrode may be removed so that it is easier to work on. Drill a 9/64-inch hole about 3/8 inch past the bend line, and then cut the positive electrode about 3/8 inch beyond this hole (see Fig. 14-25). Before bending it, put the positive electrode back through the center tube, and then bend it at the marked bend line. When bent, the bottom of the positive electrode should be the same length as the core stand brackets so that these all rest flatly against the bottom of the PVC base (Fig. 14-22). Feel free to adjust the actual bend location as needed so that the electrode will be even with the bracket.

Secure the top of the positive electrode to the stainless tube with the one screw left free (see Fig. 14-19), and tighten all remaining screws if needed (Fig. 14-23). Because the connection point is a slot, there is some play in adjusting the height of the positive electrode such that it will rest evenly against the bottom of the PVC base.

Fabricate a negative electrode from a strip of about 3½-inch stainless steel (Fig. 14-24).

Figure 14-23   Positive electrode attachment to top of core tubes.

Form a 90-degree bend ½ inch from end. Drill a 9/64-inch hole in the center of the bent leg. Then drill a slotted hole at the end of the main section, which will allow the negative electrode to be adjusted to the height of the bottom bracket. These must be flush to the mounting plug shown later (see Fig. 14-27).

Note how the positive and negative electrodes are both at the same height as the plastic bracket (Fig. 14-25). This is important because the assembly must fit flush when mounted to the PVC base bottom.

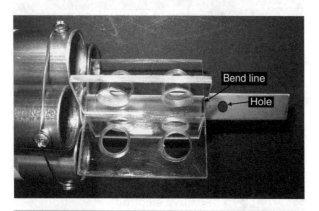

Figure 14-22   Positive electrode attachment to bottom end of assembly.

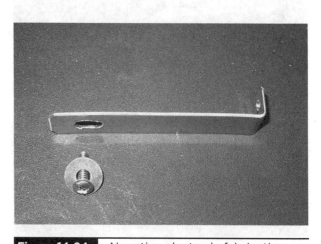

Figure 14-24   Negative electrode fabrication from 3½- × ½- × 1/16-inch stainless.

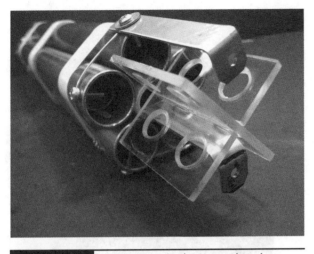

Figure 14-25 Reactor core bottom showing electrodes and alignment.

Temporarily place the bottom of this reactor core into the PVC bottom base, centering the reactor cell as best as possible in the PVC base (this is mainly for aesthetics and will not affect the cell's operation, but it's still nice to have a good-looking core); then mark where the holes of the positive and negative electrodes come to rest on the inside of the PVC bottom base, and this is where the holes will be drilled (Fig. 14-26). Remove the reactor core, and then drill 3/32-inch holes at the

marked locations, through both the PVC bottom base and the plywood stand.

Then remove the PVC base and slightly open up the holes through the plywood to 9/64 inch for clearance of the 1½-inch screws (this will give them enough clearance to be inserted without any resistance from the threads). Then, on the top of the plywood stand, countersink a ½- to ¾-inch-diameter hole approximately ¼ inch deep to allow space for the nut that will be on the bottom of the PVC (see Fig. 14-28 for reasons why).

Now goop up (use 3M super-weather strip adhesive, yellow in color) the area around the holes in the PVC base as well as the bottom of the electrodes. Try not to get too much adhesive into the hole because the metal bolts will need to pass through, and it is good to minimize the amount of adhesive on the threads at this point (Fig. 14-27). Next, run the 6-32 × 1½-inch stainless screws through the electrodes and also pass them through the PVC base (the core now can be attached to the PVC base). Now, on the bottom of the PVC base, apply some more adhesive (it is easiest to lay the assembly sideways, hold the bolts in place with your fingers or a couple of screwdrivers, and apply the adhesive to where the bolts protrude from the PVC base).

Figure 14-26 PVC bottom base with marked holes for electrodes of Fig. 14-25.

Figure 14-27 Method used for sealing mounting screws.

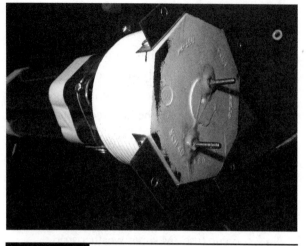

Figure 14-28 Method used for sealing bottom of mounting screws.

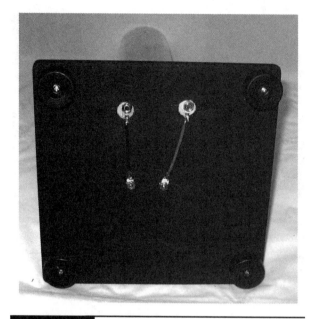

Figure 14-29 Wiring banana jacks to the wooden base.

Alternatively, you may pull on the bolt threads as you push against the PVC bottom base to keep things snug. (A helper for this step can make things easier.) After some adhesive is down, place a washer over each bolt, and thread down a nut. Once the nut is tightened up (snug but not overly tight), apply some more adhesive to cover the washer and nut (Fig. 14-28). The seal must be airtight because the assembly can be exposed to over 50 pounds of pressure. This sealing is a very important step for safe and efficient operation.

Note how the bolts and adhesive protrude from the PVC base bottom, which is why the indents were cut into the top of the plywood stand before.

Once the adhesive has dried, secure the PVC base bottom and reactor core assembly to the plywood stand. The bolts coming out of the bottom should fit through the holes because these were all drilled in a single pass (see Fig. 14-26). Run 12-gauge wire to a lug, slip this over the bolt, and secure with a nut.

Use heavy-duty lugs or build up solder lugs with excess solder because current can be 25 amps. Use 12-gauge wire and check for hot spots when the reactor cell is running. You may find that an indent is also useful on the bottom of the plywood stand to keep the circuit from grounding. In Fig. 14-29, the top connections have indents, whereas the bottom connections have no indents. If it is preferred

not to indent the center bolts coming from the core (because the top of the plywood stand is also indented, and indents on both sides weaken the connection), a small protective piece of 1/16-inch sheet plastic can be used to cover the terminal points and minimize the risk of grounding. The reactor core is now complete.

We used clear polycarbonate for the enclosure, but less-expensive nontransparent PVC may also be used. Cut the tube to 26 inches in length. Given the high pressure requirements, glue and screw caps securely. PVC glue works well, as does fiberglass resin. Also use four No. 6 self-tapping screws through each cap and into the enclosure tube, spaced 90 degrees apart. When threading, wrap with several turns of Mylar tape to prevent gas leakage.

Drill 27/64-inch holes (or whatever matches the gauge and male-male connector threads) as far apart as possible in the top cap, and tap out with a ½-20 NF tap (Fig. 14-31 and 14-32).

Use Teflon tape and standard plumbing techniques when threading together because the connections must be totally leakproof. Thread in the ball-valve assembly first. Then thread in the meter (Figs. 14-33 and 14-34).

Figure 14-30 Reactor core assembly ready for enclosing into vessel.

Figure 14-31 Main housing and slip caps.

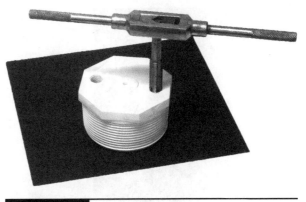

Figure 14-32 Top cap tapping.

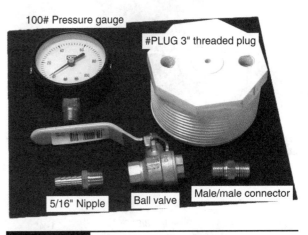

100# Pressure gauge

#PLUG 3" threaded plug

5/16" Nipple   Ball valve   Male/male connector

Figure 14-33 Top cap components.

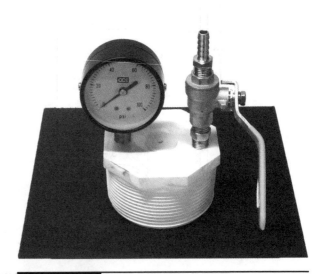

Figure 14-34 Top cap assembled.

Figure 14-35 Final photo of enclosure assembly ready to attach to reactor core assembly (Fig. 14-30).

Use Teflon tape around the top and bottom threads, wrapping two or three times (enough to be airtight but not so much that threading the pieces together requires excessive force) (Fig. 14-35). This will ensure a pressure-tight seal and will be easier to disassemble for cleaning of the reactor cell core.

Bubbler Assembly

The Bubbler (Fig. 14-36) is similar to the reactor cell, but without the internal stainless steel cell or the electrical leads running out of the container.

For the 3/8-inch pipes, use copper, aluminum, or plastic rated to 90 to 100 lb/in^2 pressure. For the body, use 3-inch Schedule 40 PVC and 3-inch ID PVC end caps.

Drill the holes in the cap slightly smaller than the tubes for a press fit. Tap the tubes into place using a hammer and wood block. Long tubes should be of a length that it will be about 2 inches from the bottom when assembled. Center tube is flush with the cap. Seal these tubes with PVC sealer or RTV/weather strip adhesive (Figs. 14-37 and 14-38).

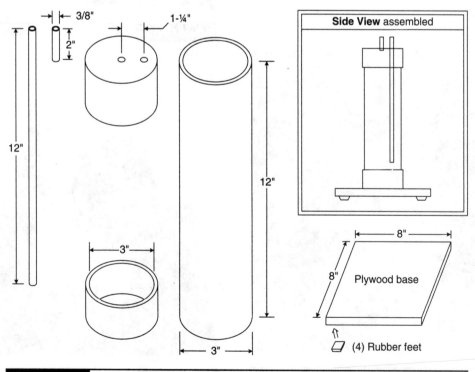

Figure 14-36 Bubbler fabrication aid sketches.

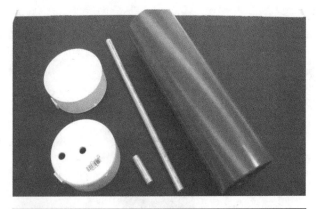

Figure 14-37   Fabricated parts for bubbler flashback protector.

Figure 14-39   Sealing end caps to the enclosure tube.

Cement PVC caps to the ends (Fig. 14-39). They must be airtight and able to withstand the operating pressure of the system.

Drill pilot holes, and screw four No. 6 × 3/8-inch sheet-metal screws into the PVC caps, putting sealant on the threads if needed (Fig. 14-40).

Attach the bubbler with adhesive to a 6- × 8- × ½-inch plywood base. Attach four stick-on rubber feet to bottom surface (Figs. 14-41 through 14-43).

The power conditioner (HHOPC10, see Chap. 13) produces current pulses and has a temperature feedback loop that provides high output efficiency. HHO gas production can be achieved with different power supplies, but production will be much slower.

The reactor's HHO gas output must be sent through the bubbler(s) to prevent flashback and a dangerous explosion. The input hose to the bubbler goes to the longer tube immersed in water. The output from the bubbler is from the short tube, which then connects to the HHO device (your project).

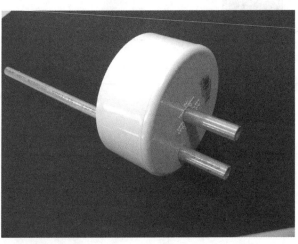

Figure 14-38   Cap with input and output tubes.

Figure 14-40   Sealing end caps to the enclosure tube with additional screws.

Figure 14-41 Base-plate assembly.

Figure 14-42 Bubbler final assembly.

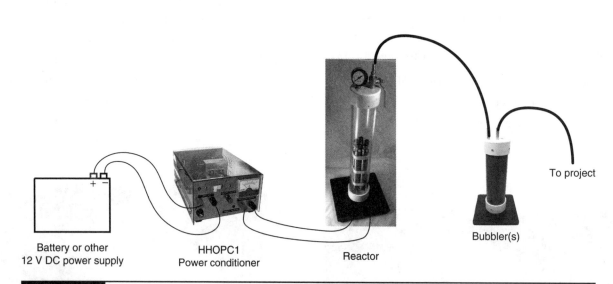

Battery or other
12 V DC power supply

HHOPC1
Power conditioner

Reactor

Bubbler(s)

To project

Figure 14-43 Schematic of final setup.

Part Info.	Quantity	Description	Part #
		HHO Reactor Cell	
ENC1HHO	1	Clear cell enclosure 3.5 × 20 inch LEXAN tubing fab You may use lower-cost Schedule 40 PVC	ENC1HHO
Ladders	2	½- × ½- × 18-inch stainless steel strips	
TUBESS1	5	Tubes stainless 1-inch OD × 0.065-inch wall × 11-inch No. 316 fabrication	TUBESS1-316
TUBESS75	5/7	Tubes stainless 0.75-inch OD × 0.065-inch wall × 11-inch No. 316 fabrication	TUBESS75-316
PLUG31829	2	3-inch threaded plug, Schedule 40	
ADPT30330	2	3-inch thread to 4-inch slip	
TUBEGT78	1	1-inch OD × 12-inch fiberglass tubing	
	15	11-inch-length nylon tie wraps	
BB1	1	1- × 1½-inch No. 22 galvanized steel for base brackets	
	16	6-32 "sheetmetal" screws	
	5	6-32 × 3/8-inch nylon screws	
	10	6-32 × 3/8-inch stainless steel screws	
		2-feet of no. 18 solid stainless steel buss wire	
	1	¾-inch thick by 12-inch square plywood base	
	1	100-lb pressure gauge	
	1	Brass ball valve	
	1	5/16-inch brass nipple	
	1	Male-male brass connector	
	4	Rubber feet for bottom of base	
		Optional assembled reactor core only (not the complete reactor cell, just the core)	HHOCORE10
		Bubbler	
		14-inches of 3/8-inch pipe (copper, aluminum, or plastic)	
		12-inches of 3-inch OD PVC tube	
	2	3-inch ID PVC end caps	
	8	6-32 "sheetmetal" screws	
	1	¾-inch thick by 8-inch square plywood base	
	4	Rubber feet for bottom of base	

TABLE 14-1 Parts List*

*Most parts should be available through electronics or hardware stores, but those more difficult to acquire are listed with a "Part #" and are available through www.amazing1.com if needed.

CHAPTER 15

HHO Dry Cell

Overview

This HHO Dry Cell (Fig. 15-1) is a variation on producing hydrogen gas from the Chap. 14 Reactor Cell. We call it a *dry cell* because the electrified plates are not entirely immersed in water and therefore require a lower amperage to generate HHO gas. It can be made more easily with a greater surface area and thus is capable of producing HHO gas faster. It is also easier to build but has a lower maximum pressure output (which can be seen as a safety feature—however, HHO gas should *never* be considered safe, so exercise extreme caution when using and producing HHO gas, such as using multiple blowback protectors, working with only small volumes and low pressures of HHO gas, and *never* storing HHO gas).

Hazards

All the hazards, dangers and warnings of Chap. 14 apply equally here. Be extremely careful when generating and working with HHO gas. **Eye protection should be worn when making, testing, and operating this device.**

Difficulty

Intermediate. Mechanical cutting, drilling, and tapping, with no soldering or circuitry required.

Tools

Drill, saw (handsaw, jigsaw, or band saw), sharp knife, metal cutter or metal saw, hole tap.

Assembly

First, cut seven rubber dividers of 1/8 inch thick to size; the example shown here is a 6- × 6-inch sheet (Fig. 15-2), but larger sheets can be used if a larger cell is desired (e.g., 6 × 10 inches for the Cell in Fig. 15-11). These rubber sheets are easy to cut: A sturdy scissors or drywall knife should suffice. As always, be careful when using cutting instruments, and make sure that the rubber dividers are held well in place before cutting.

Figure 15-1 HHO dry cell.

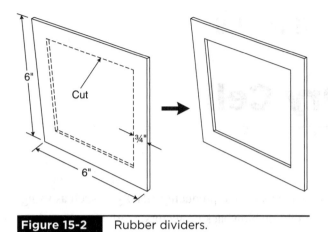

Figure 15-2 Rubber dividers.

Then fabricate two of the heavy ½- or ¾-inch-thick plastic outer sheets. See if your plastic supplier can cut these to 8 × 8 inches in size. If not, and you must cut these sheets yourself, you will find that the plastic easily overheats from a saw blade (such as a jigsaw) and will bind up in seconds while turning the cutting edge into a mushy blob of melted plastic. An excellent solution to this is to fill a standard 5-gallon bucket with water to about ¼ inch from the top. Then rest the piece of plastic being cut over the bucket such that the jigsaw blade "licks" into the water with each stroke. This will keep the blade and the surrounding plastic wet and cool and prevent it from overheating while not drawing up so much water as to risk electrically shorting or damaging the saw (or you). (Do not run a hose over the top because this introduces too much water on the top of the plastic sheet that tends to go everywhere except on the blade where it's needed. Simply having the blade lick into the water will provide all the cooling required.) Do be careful when doing this, however, because you will be operating 115 Vac near a bucket of water! Be certain that the ground surrounding the bucket is bone dry. And be *absolutely* certain that you are running the saw from a ground fault circuit interrupter (GFCI) wall socket. If you do not have a GFCI socket, then get a GFCI adapter for $20 from a local hardware store that plugs into a standard wall outlet and gives GFCI protection. This is crucial and should not be skipped because any

water short will immediately trip the GFCI fuse and stop the flow of electricity. Needless to say, we do not take any responsibility for the outcome of your methods, and do not attempt this unless you have considerable experience using jigsaws and are of sufficient strength to hold the plastic sheets steady as they are cut.

Lexan is more expensive than polyvinyl chloride (PVC), but it is good to use at least one sheet of clear Lexan, which allows seeing into the cell to verify that everything is working properly, as well as to judge when the cell should be cleaned.

Drill 14 holes as shown in Fig. 15-3, with diameters of 9/32 inch to accept the ¼-inch bolts. Two of these sheets need to be made, so clamping them together and drilling through both will ensure that the holes are aligned. A drill press works best here to both keep the hole straight and to control the rate of drilling so as not to overheat the plastic and cause it to melt or glom up. A slow trickle of

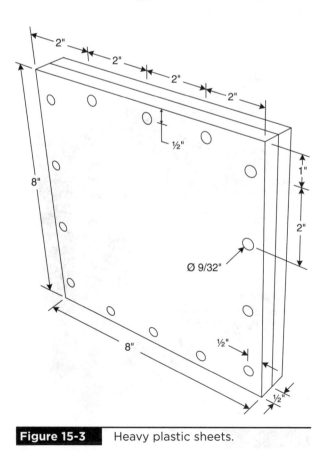

Figure 15-3 Heavy plastic sheets.

water over the hole and drill bit helps tremendously to keep the plastic cool, if you don't mind a slightly wet workspace (towels help); as always, run such a drill from a GFCI outlet, although the water should stay on the plastic and run off the sides and have no risk of contacting the electrical wiring—but, nonetheless, always be careful. Also, set the drill to a slower speed to further reduce overheating. These 14 outside holes should be centered back about ½ inch from the edge.

Then drill and tap two holes in the rear sheet for the bottom water input and top gas output connections (Fig. 15-4). In the build shown here, the holes are tapped for a ¼-inch nipple with ½-inch National Pipe Threads (NPTs), but your holes, of course, should be tapped according to the size of your connectors.

Next, cut six stainless steel sheets 1/16 inch thick (Fig. 15-5). Make two of these sheets with a block at the top about 1 × 1 inch in size, which will serve as the electrical lead for the positive (+) and negative (−) terminals. Drill and tap a hole

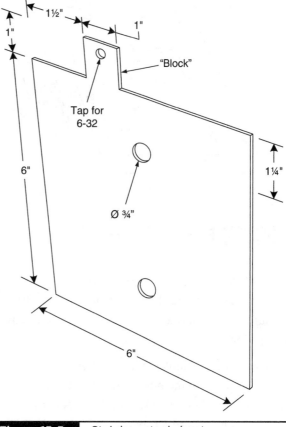

Figure 15-5   Stainless steel sheet.

into this block for a 6-32 screw (or whatever size will be used to secure the electrical terminal). The other four stainless steel sheets do not need this block. Locate the block such that it avoids the bolts holding the cell together (the offset given in Fig. 15-5 for block placement is an estimate); if this can't be done for size reasons, or if it is forgotten, then drill larger holes through the blocks to allow the bolts to pass through with some clearance (at least a 3/8-inch hole for a ¼-inch bolt), and insulate the bolts with a sleeve of rubber tubing to prevent contact with the block (see Fig. 15-10 for an example of this). Also drill two larger holes about ¾ inch in diameter (the tap in the back plastic end plate is ½ inch, but make the hole in the stainless sheets ¾ inch), near the top and bottom of each stainless steel sheet, to allow water and gas to circulate between the plates.

Next, place the heavy ½-inch back sheet down on a table, and center a rubber divider on it so that

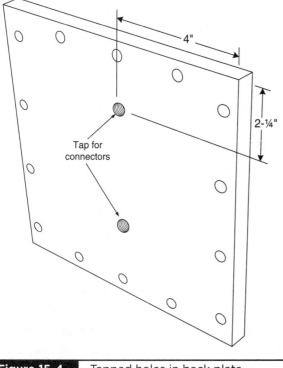

Figure 15-4   Tapped holes in back plate.

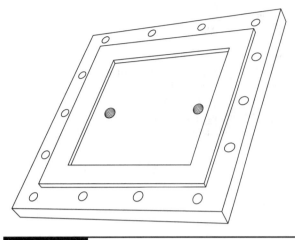

Figure 15-6 Back plate with rubber divider.

the borders are about 1 inch per side (Fig. 15-6). Ensure that the surfaces are clean and clear of any debris for a good watertight contact.

Then stack a stainless plate on top of this (one of the plates with the block attachment), and keep stacking and alternating rubber dividers with stainless plates until all six stainless plates and seven dividers are stacked together. Be sure that the last stainless sheet also has a block at one edge, flipped over so that the blocks are on opposite sides (Fig. 15-7). Screw an electrical terminal to each block, and note that the screws will likely need to

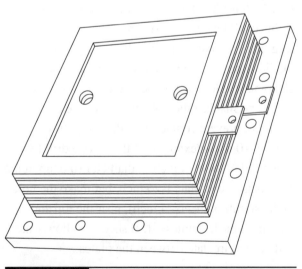

Figure 15-7 Stacked rubber dividers and steel plates.

be filed down in order to fit into the 1/8-inch gap (see Fig. 15-10 for the reason why).

Then place the heavy ½-inch front plate on top of this stack, and hold the whole thing together with a couple of wood clamps or C-clamps (if the clamps are metal, use paper to protect the plastic from scratches). Now the cell can be stood up vertically, and ¼-inch bolts and washers can be put through each of the 14 outer holes, with corresponding nuts and washers to hold against the back plate.

When tightening the nuts, start in a corner, lightly hand-tighten the nut, then move to the opposite corner and lightly hand-tighten, and then finish off the sequence (Fig. 15-8, Step 1). After the corners are finger-tightened, tighten the remaining bolts with a similar "oppositional" pattern (Fig. 15-8, Step 2). Retrace these two patterns with a wrench, tightening no more than one-quarter turn each time. Keep repeating these two patterns until the bolts start to firm up. If this seems like a complex pattern at first, remember that it is only moving from one bolt to its farthest "opposite" bolt and tightening every bolt only a little bit each time so as to equalize the compression as the plates are squeezed together. The idea is to gradually compress the cell evenly across its surface and not put too much compression on one side compared with another, which may cause the rubber seals to distort or form air gaps or weak spots. Do not overtighten the bolts or the plastic will break. As a rough guide, about 10 ft-lb of torque should do. If it leaks, snug up the bolts a little bit more.

The dry cell is now nearly finished and should resemble Fig. 15-1. Note that in this image there is a terminal block on each plate, and the bolts are going through the stainless steel terminal blocks (an early design). The dry cell actually needs only two terminals (one on the front plate and one on the rear plate), and the terminals can be positioned so that they do not touch the bolts (such as shown in Fig. 15-7). Apart from these differences, your dry cell should look like Fig. 15-1.

Step 1

Step 2

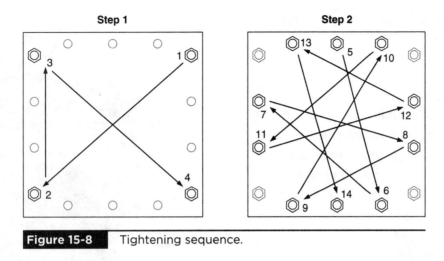

Figure 15-8 Tightening sequence.

Wrap the plumbing attachment threads with several layers of Mylar plumbing tape, and screw them into the back plate (Fig. 15-9). Note that these can be straight-through or elbow-angle attachments, whatever your preference or needs, because both work equally well.

The electrical terminals have been secured to a stainless plate through a tapped hole (see Fig. 15-5). Note how the screw securing the electrical terminal has had its length reduced in size to a ¼-inch stem (or less); otherwise, it would interfere with the cover plates (Fig. 15-10).

The dry cell is now complete and ready to generate HHO gas. See Chap. 13 for how to make the Power Conditioner and the last few pages of Chap. 14 (Figs. 14-36 through 14-43) for how to make the Bubbler assembly that prevents flashback

Figure 15-9 HHO dry cell (reverse view).

Figure 15-10 HHO dry cell terminal detail.

into the HHO dry cell and an overview schematic for the final setup.

Optional Sizes

This dry cell can be made in larger sizes and with more plates, but the input amperage will need to be stepped up accordingly to provide an increased HHO gas output. Figures 15-11 through 15-13 show a dry cell with nine plates of 6- × 10-inch dimensions, effectively giving 283 percent more surface area than six plates of 6- × 6-inch outer area (remember the ¾-inch rubber seal). Note that more bolts are used, of course, to hold the cell together. And also note that the electrical terminals are done properly here, with one on the front plate and one on the rear plate, each of them spaced away from the retaining bolts.

Note the changes in the rear plumbing connections: The bottom connection is lower water, and the middle

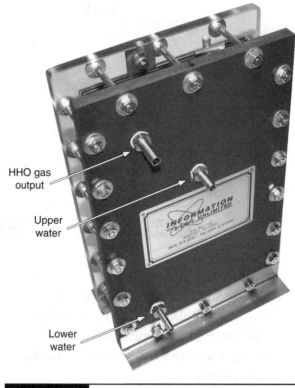

Figure 15-12   Larger HHO dry cell, rear.

connection is upper water; the water will circulate with the water supply container once the power conditioner is turned on and HHO gas is generated under pressure in the Dry Cell. The top connection is for the HHO gas output.

Also note the nine plates used in this dry cell, providing a greater surface area to generate HHO gas (Fig. 15-13).

Setup

When finished, the dry cell is part of a system that generates HHO gas consisting of four components:

- 12-Vdc power source (such as a car battery or a 12-Vdc power supply with a decent amperage—at least 7 amps; note that a battery *charger* should *not* be used because 12-V battery chargers have an actual output of 14 up to 20 Vdc and may overload and burn out the power conditioner, which is made to run on 12 Vdc)

Figure 15-11   Larger HHO dry cell, front.

- power conditioner (from Chap. 13)
- Cell (such as the dry cell here)
- One or two bubblers (or antiblowback containers from the last pages of Chap. 14)

These components are then connected together to produce HHO gas (Fig. 15-14).

First, the power conditioner is connected to the dry cell, which is kept full of water by the water supply (Fig. 15-15).

Then the HHO gas flows through bubblers—two are used here to be extra sure in preventing any blowback (Fig. 15-16). Also note how the HHOIGNITOR (Chap. 16) has been attached to the exit tube, with the negative output connected to the copper tube and the positive high-voltage (HV) wire positioned to create a spark at the exit point and ignite the HHO gas as it leaves. The spark needs to be fired only once, after which the flame will burn as long as HHO gas is produced.

When running, the whole setup will look like Fig. 15-17, with the dry cell frothing up bubbles of HHO gas, which then goes through the water

Figure 15-13 Larger HHO dry cell showing nine plates.

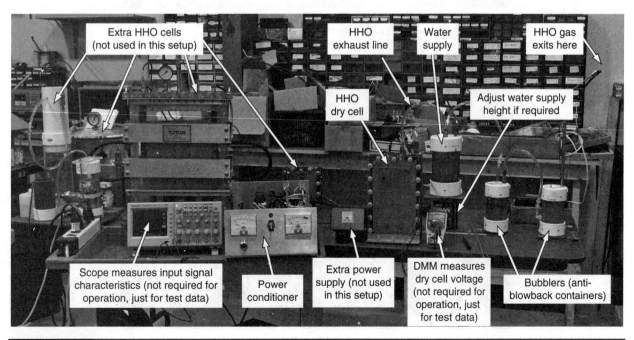

Extra HHO cells (not used in this setup)

HHO exhaust line

Water supply

HHO gas exits here

HHO dry cell

Adjust water supply height if required

Scope measures input signal characteristics (not required for operation, just for test data)

Power conditioner

Extra power supply (not used in this setup)

DMM measures dry cell voltage (not required for operation, just for test data)

Bubblers (anti-blowback containers)

Figure 15-14 Final setup.

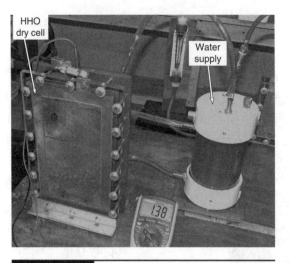

Figure 15-15 Dry cell and water supply.

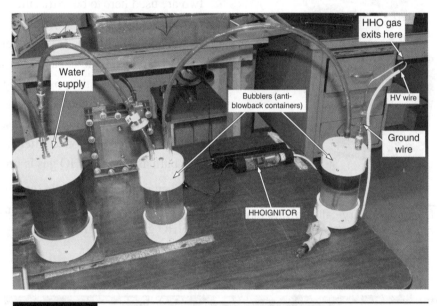

Figure 15-16 Bubblers and igniter.

Figure 15-17 HHO gas production and ignition.

supply and two bubblers and finally gas exits where it is ignited by the HHOIGNITOR (right side of Fig. 15-17).

You will probably want to use HHO gas for other purposes than just burning it as it is produced, but it is shown here so that the final full operation can be seen as to how water is converted into a flammable HHO gas. This is a hot flame, capable of reaching or exceeding 4000°F at its hottest point.

Quantity	Description	Part #
TABLE 15-1 Dry Cell Parts List*		
2	½-inch thick plastic sheets 8 × 8 inches (size determined by actual cell)	
14	Bolt, ¼ × 2½ inches (or sized to fit); quantity determined by cell size	
14	Nuts, ¼ inch; quantity as above	
28	Washers, ¼ inch; quantity as above	
	Teflon plumbing tape	
6	1/16-inch-thick stainless steel sheets 6 × 6 inches, cut & hole punched	HHODRYCORE66
7	1/8-inch-thick rubber sheets 6 × 6 inches	
	Optional 1/16-inch-thick stainless sheets 6 × 10 inches for larger HHO cell, quantity of 9 plates (be sure to adjust the size and quantity of all other components to match the increased cell size)	HHODRYCORE610
2	Electrical terminals	
2	6-32 screws, filed to length for end-plate terminals	
2–3	½-inch plumbing connectors (called *brass barb male adapters*)	

*Most parts should be available through electronics or hardware stores, but those more difficult to acquire are listed with a "Part #" and are available through www.amazing1.com if needed.

HHO Igniter/Gas Igniter

Overview

A remotely operated high-voltage igniter powered by a 9-V battery (Fig. 16-1).

Hazards

Nothing significant. The high-voltage (HV) spark produced is of low enough current to not pose any safety hazard, but it will cause an irritating and mildly painful burn. Eye protection should be worn when making, testing, and operating this device.

Difficulty

Intermediate. Familiarity with soldering.

Tools

Basic construction and electrical tools.

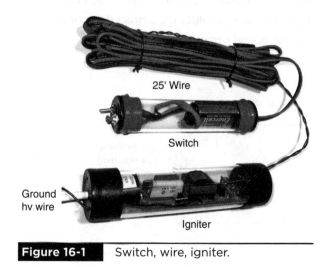

25' Wire

Switch

Ground hv wire

Igniter

Figure 16-1   Switch, wire, igniter.

Circuit Description of Pulsed Shocker

Figure 16-6 is a schematic that shows a high-frequency self-oscillating inverter circuit comprised of switching transistor Q1 and step-up transformer T1 that produces high voltage and high frequency at the secondary winding. This HV alternating current (ac) is rectified by diode D1 and charges up storage capacitor C3 or C4 through isolation resistor R3. When this voltage charges up to the breakdown potential of silicon diode for alternating current (SIDAC) SID1, the energy stored in the capacitors is dumped into the primary of HV pulse transformer T2, producing a HV pulse at the output terminals.

The oscillator circuit uses a winding on T1 to produce the necessary positive feedback to the base of Q1 to sustain oscillation. Resistor R1 initiates Q1 turn-on, whereas resistor R2 and C2 control the base current and operating point.

A charging circuit consisting of current-limit resistor R4 and rectifier diode D2 allows external charging of battery B1 when nickel-cadmium or other rechargeable batteries are used.

Switch Assembly

The Switch is a relatively simple assembly, as seen from the schematic in Fig. 16-2.

First, drill three holes in a 1½-inch metal end cap, and through these holes secure the toggle switch, push button, and light-emitting diode (LED) (Fig. 16-3).

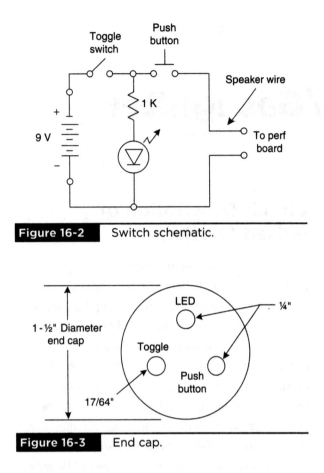

Figure 16-2 Switch schematic.

Figure 16-3 End cap.

and solder the wire, the 9-V battery clip, and the end-cap components together as per the schematic in Fig. 16-2. For ease of assembly, the components should be soldered together outside the tube (Fig. 16-5).

Then slide this circuit into the 1½- × 5½-inch plastic tube, slide the bottom end cap over the bottom of the tube, and wrap the end caps with tape to hold them in place (as seen in Fig. 16-1). The Remote Switch is now complete.

Igniter Assembly

The Remote Switch is then attached to the igniter, where the 25 ft of speaker wire can be either soldered directly to the circuit or attached with a jack (Fig. 16-6).

The igniter can be built using a *printed circuit board* (PCB; see parts list at end of chapter), or you may use the more challenging *perforated circuit board*. The perf-board approach is more challenging because the component leads must be created and routed as conductive metal traces on the back of the perf board. If using this method, we suggest that you trace the connection lines on the perf board before inserting and soldering the components. Start from a corner and proceed from left to right. Note that the perf board is the preferred approach for science projects because the system looks more homemade. The PCB only requires that you solder the correct component into the respectively marked holes.

Then drill a single hole through the other 1½-inch metal end cap and insert a LED grommet (through which the speaker wire will go and be protected from chafing). Then cut some 1½-inch-diameter tube to a length of about 5½ inches. When assembled, this will be the basic housing of the remote switch (Fig. 16-4).

Then pass a length of speaker wire (we use about 25 ft) through the bottom cap and tube,

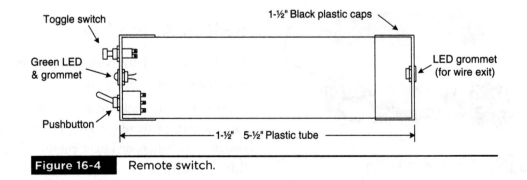

Figure 16-4 Remote switch.

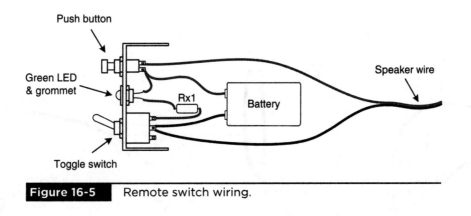

Figure 16-5 Remote switch wiring.

First, lay out and identify all parts and pieces, checking the parts list. Note that some parts sometimes may vary in value. This is acceptable because all the components are 10 to 20 percent tolerance unless otherwise noted.

Beginning hobbyists may wish to download our free *General Construction Techniques* (Item No. GCAT1) from our website before proceeding.

When you are ready to proceed, assemble the board as shown in Fig. 16-7. Start to insert

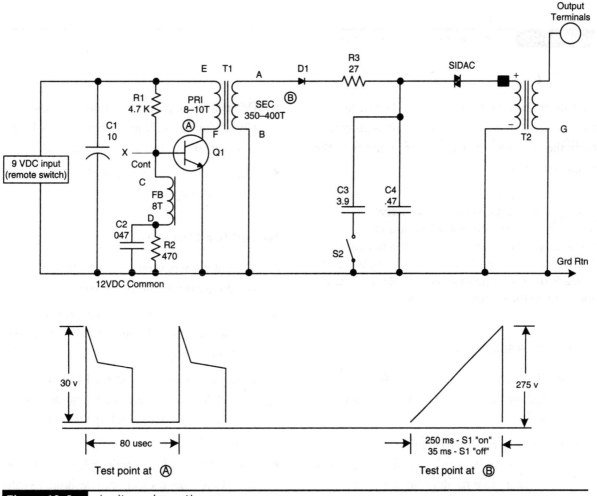

Figure 16-6 Igniter schematic.

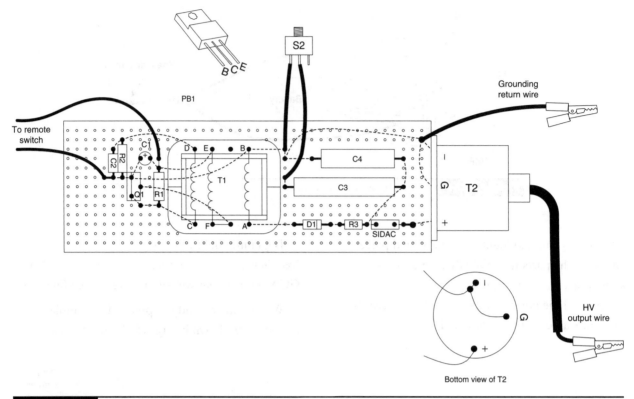

Figure 16-7   Component layout and wiring.

components into the board holes as shown. Note to start and proceed from right to left, attempting to obtain the layout as shown. Note the polarity of the components as shown.

NOTE   Certain leads of the actual components will be used for connecting points and circuit runs. Do not cut or trim them at this time. It is best to temporarily fold the leads over to secure the individual parts from falling out of the board holes for now.

Notes on Assembly

- The SIDAC may have three pins. You may disregard the center pin. The device is not polarized.

- Run a bead of RTV silicon rubber to secure T2 to the assembly board. This will keep wires from breaking as a result of flexing as the assembly is moved.

- Determine the proper position of T1 by identifying the winding designated "A,B" using an ohmmeter to measure around 30 Ω.

- The ground return wire and HV wire should be about 2 ft long.

Notes for Optional PCB

- The cut corners shown on the silk-screened Q1 is the base pin B end of the part.

- R4 and D2 shown on the schematic are mounted between J1 and S1 (for recharging the battery only).

- Disregard the silk screen for T1 on this project. You can determine the output winding by measuring approximately 9 Ω using an ohmmeter.

- Switch S1 may be a single pole single throw (SPST) for this project because it only switches the battery power.

Once the device is assembled, verify wiring, proper components, and polarity via the schematics. Check for cold solder, excess solder, and bare-wire bridge shorts. The unit is ready to test.

Testing the Unit

1. Position the bare end of the ground return lead (GRD RTN) so as to allow a ½- to ¾-inch air gap between the output pin of T2.

2. Connect 12 V from a bench power supply or use eight AA nickel-cadmium batteries at 1.25 V each or eight AA alkaline batteries at 1.5 V each for 12 V.

3. Note that with S2 open, you should note a fast pulsing action producing a thin bluish discharge. This can cause a very mild electrical shock and could be used safely within reason as a prank.

4. Note that with S2 closed, there is a thick, slow pulsing discharge. This can produce a painful shock and is intended for use against animals and so on.

5. Current draw with the unit properly operating should be approximately 250 milliamps.

6. Check power tabs of both Q1 and the SIDAC. These should be cool to warm to the touch.

7. You may verify proper operation using a scope and noting the wave shapes, as shown Fig. 16-6.

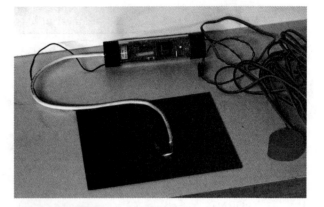

Figure 16-8 HV spark.

Notes on Pulsed Igniter

Input is shown operated with 9 Vdc. The unit, however, operates reliably within 8 to 14 Vdc.

The unit may be housed in any suitable enclosure. Shown here it is placed inside a plastic tube. You also may fill it with wax. This will allow operation in moist environments while still allowing the unit to be easily unsealed should problem or changes occur.

The igniter is now assembled. Connect the HV and ground wires to your project, and press the toggle switch to deliver a 20-kV pulsed spark (Fig. 16-8).

TABLE 16-1	Remote Switch Bill of Materials*			
Ref No.	**Quantity**	**Description**		**Part #**
		Remote Switch		
Sx1	1	SPST 3-amp push-button switch or equivalent		
Sx2	1	SPST 3-amp flip switch or equivalent		
CL1	1	Battery clip		
CAP1x, 2x	2	1½-inch metal cap		
SPKWR		25 ft no. 18 or 20 speaker wire		
LED1	1	Green led		
GRMT	2	Led gromet		
Tube	1	1½- × 5½-inch plastic tube		
Rx1	1	1-kΩ resistor		
		Optional fully assembled HHO remote switch, ready-to-use		HHOSWREM10
		Igniter		
R1		4.7-kW ¼-W carbon film resistor (YEL, PUR, RED)		
R2		470-W ¼-W carbon film resistor (YEL, PUR, BR)		
R3		27-W ¼-W carbon film resistor (RED, PUR, BLK)		
R4		100-W ¼-W carbon film resistor (BR, BLK, BR)		
C1		10-μF 25-V electrolytic capacitor vertical mount		
C2		0.047-μF 50-V polyester capacitor		
C3		3.9- TO 4-μF 350-V polyester capacitor		
C4		0.47-μF 250-V polyester capacitor		
Q1		MJE3055 *NPN* transistor TO220		
D1, 2	2	IN4007 1-kV rectifier diode		
SIDAC		300-V sidactor switch sidac		SIDAC
T1		Switching square wave transformer 400 V		TYPE1PC
T2	1	25-kV pulse transformer		CD25B
PB1		5- × 1.5-inch grid perf circuit board		
WR20B		36 inches no. 20 vinyl stranded hookup wire, black		
WR20R		36 inches no. 20 vinyl stranded hookup wire, red		
WRHV20		12 inches 20-kV wire		
CAP1,2	2	1-5/8-inch plastic cap		
J1	1	3.5-mm mono jack		
12DC/.3		Optional 12-Vdc 0.3-amp wall adapter		12DC/.3
		Optional printed circuit board		PCLITE
		Optional fully assembled HHO ignitor, ready-to-use		HHOIGNITOR10

*Most parts should be available through electronics or hardware stores, but those more difficult to acquire are listed with a "Part #" and are available through www.amazing1.com if needed.

CHAPTER 17

HHO Bomb

Overview

The basic concept is simple: Fill a bag with HHO gas, and ignite it for a fun, percussionary event. But the details are important for a controlled and safe detonation without causing any backflash damage to the HHO gas generator.

Hazards

Building this device entails no hazards, but working with HHO gas is very dangerous, and all applicable safety precautions must be followed. **Eye protection should be worn when operating this device.**

Difficulty

Simple.

Tools

Wire stripper and heavy duty scissors/knife for cutting small plastic tubes.

Assembly

Cut a 6-inch length of 3/8-inch clear plastic tubing, and drill a 1/8-inch hole approximately 1¼ inches from one end. Also cut a 6-inch length of 3/8-inch-diameter flexible rubber hose (Fig. 17-1).

Cut an 18-inch length of 24-gauge speaker wire (or longer, up to several feet, if you want more safety when attaching igniter; see Fig. 17-9), and strip only one end. Then feed the wire through the

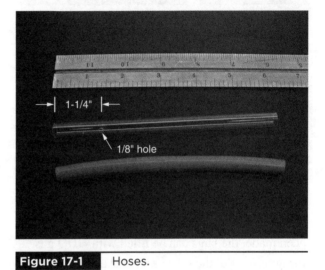

Figure 17-1   Hoses.

1/8-inch hole until it extends about 1½ to 2 inches from the far end of the tube (Fig. 17-2).

Apply silicone to seal around the hole and wire (to prevent HHO gas from escaping), but only seal this outer hole—do *not* apply so much silicone as to block the flow of HHO gas through the tube (Fig. 17-3).

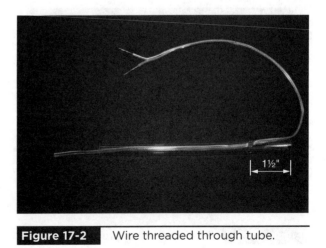

Figure 17-2   Wire threaded through tube.

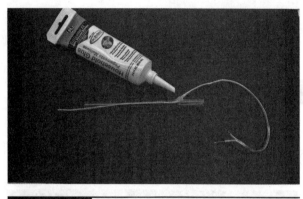

Figure 17-3 Sealing the tube hole.

Fold the wire, and then tape over the silicone and hole (equal parts tape on both sides of hole) (Figure 17-4). *This must be an airtight seal!*

Push ½ inch of the rubber hose onto end of the plastic tube, and then attach a crimp clamp to the rubber hose (Fig. 17-5).

Now that the reusable feed and ignition tube has been made, the bags can be filled with HHO gas for ignition. Put the tube into a plastic bag, with the unstripped end of the wire extending into the bag (Fig. 17-6). Remove as much air as possible from the bag.

NOTE Warning: HHO gas is very volatile. Although rare and requiring the right conditions, a good static shock (anything above 20 µJ) can ignite the mixture.

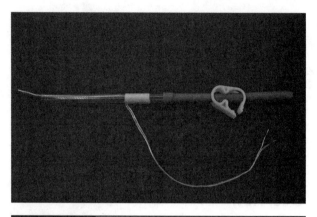

Figure 17-5 Rubber hose attached with crimp clamp.

Before filling the bag with HHO gas, seal it around the tube with an elastic band (Fig. 17-7).

Filling and Detonating

1. Generate 20 to 30 pounds of HHO gas in a reactor cell.

2. Connect the rubber feed hose to the reactor cell output.

3. Open the reactor cell's ball valve to fill the bag (should take a few seconds).

4. When the bag is full, compress the crimp clamp on the feed hose, and then detach the feed hose from the reactor cell (Fig. 17-8).

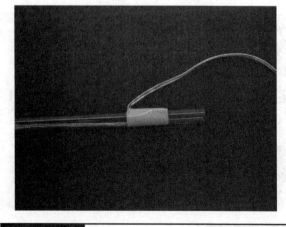

Figure 17-4 Hole further secured with tape.

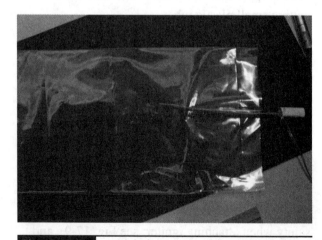

Figure 17-6 Initial placement of feed tube in bag.

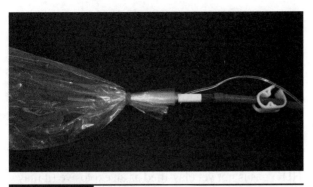

Figure 17-7   Elastic band to seal the bag.

(Remember, you need to have a short length of hose or you'll be pushing lots of air into the bag before the HHO gas arrives. A short hose means that the bag will be filled with HHO gas while it's near the reactor cell. Once the bag is filled, be sure to detach the hose from the reactor cell and move it away before igniting! If you fire this device while it's attached to the reactor cell, you run the risk of damaging the cell from the explosion and another risk from

the backflame propagating up the tube and into the reactor cell, causing a secondary explosion inside the reactor cell.)

5. Move the bag-tube combo away from the reactor cell to a safe distance. Handle with care because a jolt of static electricity still can set this off (people in arid environments, take note). It is unlikely that this will happen, but carry the device by the tube, and wear heavy leather gloves and hearing and eye protection just in case!

6. Connect the stripped end of the wire to a remote igniter with alligator clips (Fig. 17-9). You will need a high-voltage spark for igniting (see our HHOIGNITOR10 in Chap. 16).

7. Back at the igniter switch, turn on the safety, and press the button. Boom (Fig. 17-10)!

A fun thing to do is calculate the amount of water involved in one of these detonations.

Density of HHO = 0.000535 g/cm^3 (H_2O in gaseous form).

Density of water = 1 g/cm^3 (H_2O in liquid form).

A drop of water = 0.05 to 0.1 g.

The plastic bag used in this ignition could be considered a cylinder of about 18 centimeters (7 inches) in length and 9 centimeters (3.5 inches)

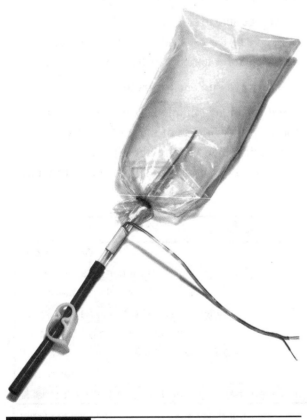

Figure 17-8   Bag filled with HHO gas.

Figure 17-9   Bag filled with HHO gas and attached to remote igniter.

Figure 17-10 Ignition of HHO gas.

in diameter. With the volume of a cylinder being $L\pi r^2$, this comes to roughly 1200 cubic centimeters, or 0.6 grams, of HHO gas. In other words, the power in this explosion comes from about 6 drops of water.

A neat way to demonstrate this is to have an eyedropper, before the explosion, and say to people, "The amount of energy in this explosion comes from 6 drops of water," while you release 6 drops of water onto a table, let that visual image set in for a few seconds, and then ignite the HHO gas. Then offer them a glass of water to drink! People won't look at water the same again.

Legal Notes

Given the climate of the world today, it would be wise to contact your local fire department, police department, and neighbors to let them know that you will be doing some experimenting with hydrogen gas and that they may hear some bangs as a result and so not to be alarmed. It also would be wise to check your local state and town laws (at a local library or search online because many states and municipalities are now posting their Revised Statutes Annotated (RSAs or laws) online for access to all, to be sure that HHO gas is not yet classified as an explosive (it may even be classified by amount or how it is used). If it is, you will need an explosives permit to use this gas, and proceeding with this experiment without such a permit would be unwise because you may end up with a charge and a court date (or worse).

This state of affairs from our nation's political and legal climate discourages budding scientists from performing these and other fun and exciting experiments that would pique the interest of young minds, potentially leading to a scientist or an engineer. But it's also the reality of our world, so please check your local laws and with law enforcement to be sure that what you're doing won't get you in trouble.

Also be sure to clear the area of any flammable objects (e.g., leaves, dead grass, paper, etc.), and have a hose and/or water buckets on hand, just in case.

Wear eye and ear protection!

Information Unlimited assumes no responsibility for the use or misuse of these plans. If you build this device, then you are agreeing to accept full responsibility for any and all outcomes of its use.

TABLE 17-1	Parts List*	
Ref. No.	**Quantity**	**Description**
	1	3/8-inch semi-rigid clear plastic tubing
	1	3/8-inch flexible rubber hose
	1	18-inches 24 gauge speaker wire
	1	Crimp clamp for 3/8-inch rubber hose
	1	Silicone sealant
		Rubber band
		Plastic bag—size and quantity determined by you, but start small!

*Most parts should be available through electronics or hardware stores, but those more difficult to acquire are listed with a "Part #" and are available through www.amazing1.com if needed.

CHAPTER 18

Hydrogen Pistol

Overview

This project makes a handheld pistol (Fig. 18-1) capable of launching a small projectile at moderate velocity using ordinary water as its source. It is a fun demonstration of the power of hydrogen gas.

Hazards

Eye protection must be worn when operating this device.

Difficulty

Simple to intermediate. The fabrication is relatively simple, but there is a fair amount of detail work involved.

Tools

Saw, drill, sander, tap.

 DANGER Using HHO gas under pressure is very dangerous.

WARNING These plans deal with and involve subject matter and the use of materials and substances that may be hazardous to health and life. Do not attempt to implement or use the information contained herein unless you are experienced and skilled with respect to such subject matter, materials, and substances. Neither the publisher nor the author makes any representations as to the accuracy of the information contained herein and disclaim any liability for damages or injuries, whether caused by or resulting from inaccuracies of the information, misinterpretations of the directions, misapplication of the information, or otherwise.

Assembly

First, cut several tube lengths of different diameter polyvinyl chloride (PVC):

Barrel: 15-inch length of ¾-inch schedule 80 PVC (has an inside diameter of 11/16 inch)

Coupler: 2¾-inch length of schedule 80 PVC (has an actual 1½-inch outside diameter)

Spacer: 2-inch length of a 1/16-inch-thick PVC sleeve (with an actual 1½-inch inside diameter)

Combustion chamber: 12½-inch length of 1½-inch schedule 40 PVC

Chamber sleeve: 11-inch length of 2-inch schedule 80 PVC

Handle: 7½-inch length of 1½-inch schedule 40 PVC

Make an inner cap by cutting two circles from a sheet of ¼-inch-thick PVC: one at 1-7/8-inch diameter and the other at 1-9/16-inch diameter (Fig. 18-2). Then make sure that the smaller cap will fit inside the 1½-inch schedule 40 PVC

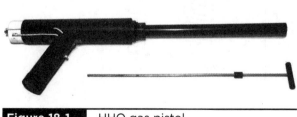

Figure 18-1 HHO gas pistol.

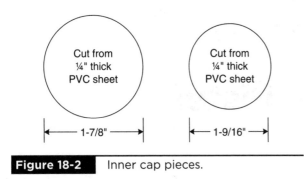

Figure 18-2 Inner cap pieces.

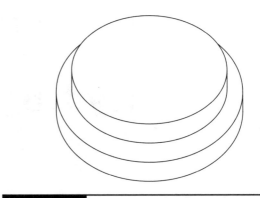

Figure 18-3   Inner cap glued together.

combustion chamber and the larger cap will fit inside the 1½-inch schedule 40 PVC end cap (see Fig. 18-12; if not, then file the edges to fit).

Next, use PVC cement to hold these two caps together, centering the small cap on the large cap. An excess of PVC cement does not need to be used here, just enough to hold the two disks. Basically just apply a dime-sized bead to the center, and push the two pieces together, sliding them around until they are well centered, and then hold them in place to set properly. Note that PVC cement sets rather fast (within 30 seconds or less), so be sure to center the disks quickly. The two disks now make up the inner cap (Fig. 18-3).

Give the inner cap a few minutes to fully dry, and then drill the center and tap it for a tire stem. Wrap the threads of the tire stem several times with Mylar plumbing tape, and screw the stem into the inner cap, being sure that the stem end protrudes from the larger disk (Fig. 18-4).

Place this assembly in the combustion chamber (with the tire stem facing outward), mark the alignment with a felt-tip pen (or anything that works on PVC), and file off any of the larger disk that overhangs the combustion chamber; otherwise, it will be difficult to slide the outer end cap over this assembly.

Next, drill a 5/8-inch hole through the center of the 1½-inch Schedule 40 PVC outer end cap, and slide this over the inner cap as it is seated in the inner chamber (Fig. 18-5) to ensure that this will fit together smoothly and snugly. If it does not go together easily, there is likely still some overhang of the inner cap; file down any remaining overhang until the outer end cap can slide over the combustion chamber with the inner cap inserted.

Once the outer end cap is assured to fit, the inner cap can be permanently secured to the combustion chamber. First, apply PVC cement to the lips of the

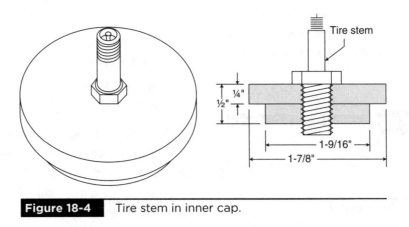

Figure 18-4   Tire stem in inner cap.

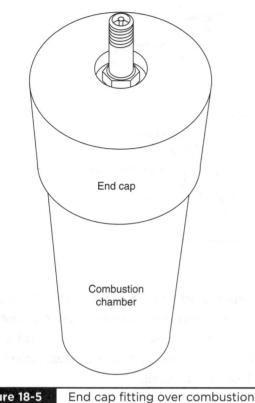

Figure 18-5   End cap fitting over combustion chamber and inner cap.

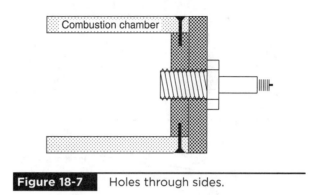

Figure 18-7   Holes through sides.

chamber into the smaller disk of the inner cap, again beveling the openings and inserting screws (Fig. 18-7).

Now coat the outside tail end of the combustion chamber and the inside of the outer end cap with PVC cement, slide them together and hold until dry. Drill and tap just one hole, this time through the side of the outer end cap, but do not penetrate the combustion chamber (set or mark your drill to not penetrate into the combustion chamber) (Fig. 18-8).

At the other end of the pistol, the barrel is made from a seamless aluminum tube inserted into 1-inch outside diameter (OD) PVC pipe (Fig. 18-9). The 14-inch aluminum tube has a 0.4375-inch OD with a 0.049-inch wall thickness and should be available in hardware and automotive stores.

Slip a 3-inch length of 5/16-inch rubber fuel line over about 2 inches of the aluminum tube.

inner cap (refer again to Fig. 18-3), and then place this back into the combustion chamber, matching up the alignment marks made before, and hold the pieces together until they set.

Then drill and tap four holes through the "bottom" of the inner cap into the combustion chamber, bevel the openings, and insert four flathead screws (Fig. 18-6).

Next, drill another four holes, staggered with the previous four, through the "sides" of the inner

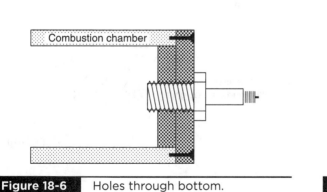

Figure 18-6   Holes through bottom.

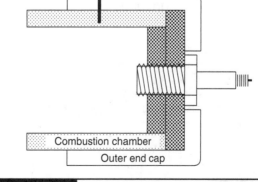

Figure 18-8   Hole through end cap sides.

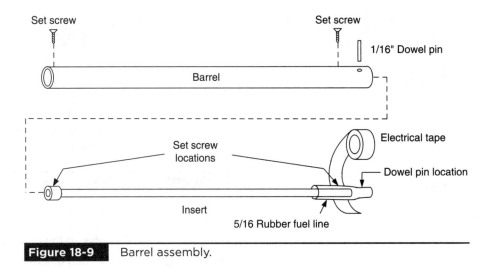

Figure 18-9   Barrel assembly.

Wrap black electrical tape at both ends of the aluminum tube (wrap it over the rubber fuel line that is itself over the aluminum tube) so that it makes a firm fit in the barrel. Fit the insert into the barrel, and drill a 1/16-inch hole about ½ inch from the inside end of the barrel, through the rubber fuel line and through the other side of the PVC. Insert a 1/16-inch metal dowel pin here, 1 inch long, to fit across the PVC barrel tube; this will serve as a physical stop for the ball-bearing projectile (and will be held in place by the coupler, but apply some epoxy or glue to keep it from falling out before that time).

Drill and tap set screws through the PVC barrel so that they secure the aluminum insert (do not puncture the aluminum!)—these should be located in the center of the electrical tape areas at the front and back of the barrel.

With the barrel, combustion chamber, and end cap assembled, the remainder of the pistol can be fit together. At each increase in the "stepping up" in size from the barrel to the combustion chamber, the mating PVC tubes should be well coated with PVC cement to ensure a good, airtight seal (Figs. 18-10 and 18-11). Start with the barrel, and work your way up in size:

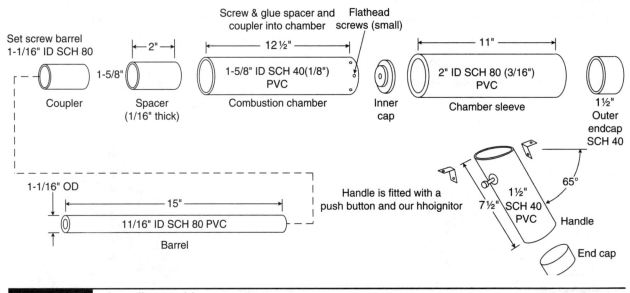

Figure 18-10   Overall assembly.

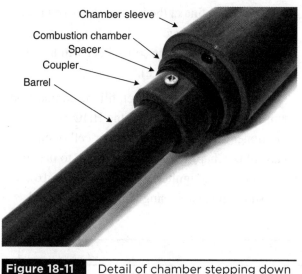

Figure 18-11 Detail of chamber stepping down to barrel.

1. Insert the barrel into the coupler, flush to the end (remember the PVC cement); then tap a set screw into the Barrel about ¼ inch from the edge of the coupler (Fig. 18-11).

2. Slide the coupler into the spacer (again, use PVC cement).

3. Push the spacer into the combustion chamber (first coat it with PVC cement), and tap another set screw (see Fig. 18-11).

4. Last, fit the chamber sleeve over the combustion chamber—this does not need any PVC cement because it is not involved with holding the barrel to the combustion chamber but only to provide support for the pressures generated in the combustion chamber.

Screws can be further used to mechanically prevent slipping, but care should be taken to avoid penetrating to the combustion area (Fig. 18-11).

Tap a hole for a stainless steel screw going into the combustion chamber that will make a spark gap with the tire valve (Fig. 18-12).

Build an HHO igniter (see Chap. 16) to fit into the handle, running the push-button switch through the handle, where it can be easily pressed with a finger, and cover the bottom of the handle with a 1½-inch plastic cap to hold in the circuitry. Secure the handle to the chamber sleeve with small L-brackets (see Figs. 18-10 and 18-13). Run HV wires to the end cap, soldering them to the stainless steel screw and tire valve (Fig. 18-13).

A charging rod also should be fabricated to load each projectile, with a "stopper" at the correct distance to prevent the projectile from being pushed too deeply into the combustion chamber

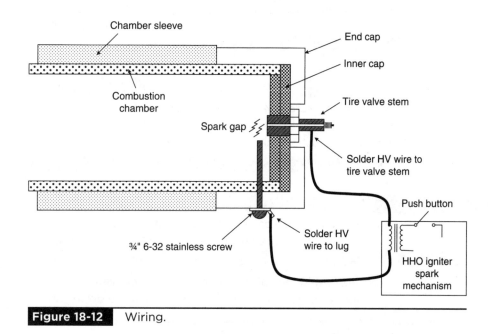

Figure 18-12 Wiring.

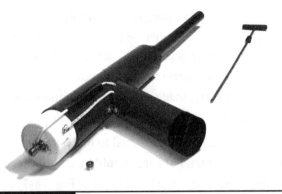

Figure 18-13 Handle and wiring detail.

(the rubber fuel line at the base of the barrel also will provide tactile feedback for when the projectile has been set properly; the soft rubber will help to hold the projectile in place).

Put a projectile in the barrel, fill the combustion chamber with HHO gas using the HHO power conditioner of Chap. 13, the reactor cell of either Chap. 14 or Chap. 15, a bubbler or two to prevent blowback, and an igniter switch (see Chap. 16). Then shoot the pistol using the power of water!

TABLE 18-1 Parts List*

Quantity	Description	Part #
	Tire Stem	
	2 ft of 20 kV high-voltage wire	WIRE20KV
	electrical tape	
	5/16-inch rubber fuel line	
	2-inch × 4-inch × ¼-inch thick PVC sheet	
	1½-inch outer endcap PVC Sch 40	
	11-inch × 2-inch inner diameter tube PVC Sch 80	
	12½-inch × 1-5/8-inch inner diameter tube PVC Sch 40	
	2-inch × 1-5/8-inch outer diameter tube PVC 1/16-inch thick	
	2½-inch × 1-1/16-inch inner diameter tube PVC Sch 80	
	15-inch × 1-1/16-inch OD (11/16-inch ID) tube PVC Sch 80	
	PVC cement	
	Aluminum tube, 14-inch × 0.4375-inch OD × 0.049-inch wall thickness	
	1-inch length × 1/16-inch diameter metal dowel pin, filed to fit flush in barrel (Fig. 18-9)	
	7½-inch × 1½-inch OD tube PVC Sch 40	
	1½-inch ID end cap	
8	Flathead screw, 7/8-inch length	
4	Flathead "set" screw, ¼-inch; file 1 to length to not penetrate combustion chamber (Fig. 18-8), 2 through insert within barrel (Fig. 18-9), and 1 through combustion chamber, spacer, and coupler (Fig. 18-11)	
1	Sheetmetal screw, through coupler and into barrel	
1	¾-inch 6-32 stainless screw	
1	Electrical lug	
2	Small brackets (these can be store-bought and bent to fit, or fabricated from 1/16-inch metal (Figure 18-10)	
1	HHO Ignitor (see Chap. 16) with push button for trigger	
	Charging rod, approx. 16-inch aluminum rod at 1/8-inch diameter for gently inserting projectile to rest against the dowel pin	

*Most parts should be available through electronics or hardware stores, but those more difficult to acquire are listed with a "Part #" and are available through www.amazing1.com if needed.

CHAPTER 19

Hydrogen Howitzer

Overview

This device (Fig. 19-1) is capable of firing a 50-millimeter projectile (Fig. 19-2) over 500 meters using ordinary tap water. The howitzer section is built from materials available from hardware stores. The design as shown was done by one of our more creative technicians and could be a museum piece. The howitzer has many extras, including a laser sight. The design of the actual barrel, chamber, and firing mechanism must be followed, but that of the remaining assembly can be at the builder's discretion. Plans include photos, schematics, and drawings to complete this fascinating and fun project.

You will need to build the power conditioner (Chap. 13), reactor core with antiblowback bubblers (Chaps. 14 or 15), and igniter (Chap. 16) to use this hydrogen howitzer.

Materials for the cell and actual howitzer are readily available. However, you may purchase them precut and ready to use from our facilities.

Hazards

A serious, deadly explosion hazard exists when using HHO gas. Because of the intrinsic danger of HHO gas, we do *not* recommend making a storage chamber to hold HHO gas, which some may be tempted to do in order to save time. *Do not make such a storage tank.* The gas is highly unstable, and the slightest internal static discharge will cause it to explode. Instead, generate HHO gas only as needed, following our stated safety precautions (such as multiple flash-stop bubblers, a limited-volume HHO chamber, pressurizing below our recommended values, etc.). **Eye protection should be worn when testing and operating this device.** Also a kinetic energy hazard, capable of serious injury, damage, and death. Treat as a firearm.

Difficulty

Intermediate for the howitzer itself (a fair amount of construction and carpentry work is required, but it is intermediate in difficulty). However, the

Figure 19-1   Hydrogen howitzer, final assembly.

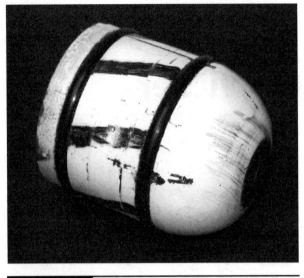

Figure 19-2   Close-up of the 50-millimeter projectile.

supporting power conditioner and igniter require more advanced circuitry work.

Tools

Saw, drill, sander, tap set, basic tools.

Theory of Operation

You will note that the 50-millimeter projectile is made with two O-rings that seal the gas from escaping up the barrel.

HHO gas is now admitted into the sealed combustion chamber at a pressure of 10 to 50 pounds. *We suggest that you start at a lower pressure and work up to the higher values, checking the integrity of the system after each firing.* We have had more than one occasion when the combustion chamber blew apart in the experimental design phase. HHO gas stored under pressure is an explosion waiting to be ignited. Read all warnings that are stated in the HHO Reactor plans and data along with use of bubbler flash-back prevention. *Take all safety precautions (eye protection, ear protection, and gloves at least).*

The projectile now moves up the barrel to a stopping screw due to the pressure of the admitted gas. Chamber pressure now builds up to the required value. The ignition system then causes the HHO gas to explode by producing a spark across the internal spark gap. The velocity of the projectile will depend on the internal gas pressure, but more does not always produce a better result. We have found that too much of an ignition pressure/charge actually will result in a slower projectile velocity. This is likely due to the materials being used (PVC tubing and a plastic projectile with rubber seals), which lose their structural integrity past a certain threshold such that the plastic bends and the rubber O-rings deform, either allowing the gas to escape or bending the plastic such that the increased friction slows the velocity. Whatever the cause, you will find a "sweet spot" of HHO pressure for the projectile's best muzzle velocity. A chronometer will greatly help in determining this.

Assembly

First, make the barrel from four pieces of polyvinyl chloride (PVC), stepping up in size to provide a larger combustion chamber (Fig. 19-3). Be sure to set the steel stop screw now (see Fig. 19-7) because this will not be possible once the barrel assembly has been fabricated.

Then make the end cap that will seal the combustion chamber (Fig. 19-4). Use PVC cement to hold these pieces together. Once dried, drill and tap as shown, including taps into the tube of the combustion chamber. Also be sure to insert the two stainless steel screws that will be the spark gap, and bend one of them so that the gap narrows from their 1¼-inch spacing to about ¼ inch at the end where the spark will occur (see Fig. 19-6). It is important to do this now because it will not be possible once the caps are secured to the combustion chamber.

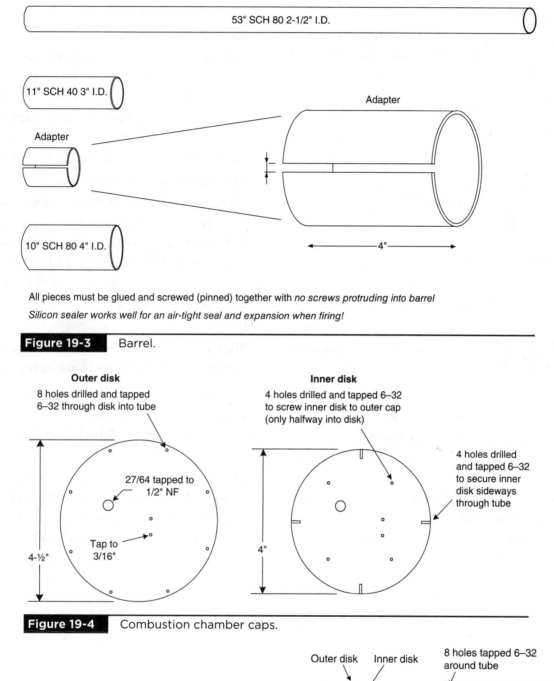

53" SCH 80 2-1/2" I.D.

11" SCH 40 3" I.D.

Adapter

Adapter

10" SCH 80 4" I.D.

4"

All pieces must be glued and screwed (pinned) together with *no screws protruding into barrel*

Silicon sealer works well for an air-tight seal and expansion when firing!

Figure 19-3   Barrel.

Outer disk

8 holes drilled and tapped
6–32 through disk into tube

27/64 tapped to
1/2" NF

Tap to
3/16"

4-½"

Inner disk

4 holes drilled and tapped 6–32
to screw inner disk to outer cap
(only halfway into disk)

4 holes drilled
and tapped 6–32
to secure inner
disk sideways
through tube

4"

Figure 19-4   Combustion chamber caps.

Secure the end cap to the combustion chamber
with PVC cement and eight screws through the outer
disk into the combustion chamber and then another
four screws through the side of the combustion
chamber into the inner disk (Fig. 19-5).

Screw the ball valve into the End Cap (Fig. 19-6).

When assembled, the overall barrel dimensions
should appear as in Fig. 19-7.

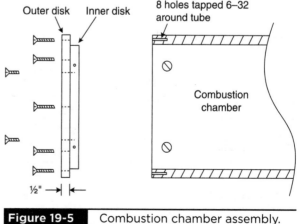

Outer disk   Inner disk

8 holes tapped 6–32
around tube

Combustion
chamber

½"

Figure 19-5   Combustion chamber assembly.

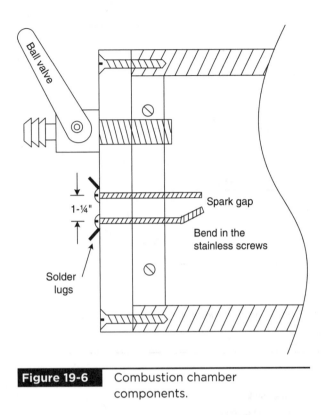

Figure 19-6   Combustion chamber components.

The projectile is a piece of plastic with two O-rings to provide a better seal when being fired (Fig. 19-8). You do not want this seal to be too tight because it might bind up in the barrel, but the right amount of tightness will help to get the maximum push from the expanding gas. We use a petroleum lubricant along the inside of the barrel to reduce friction. But be careful because the flame from the

HHO gas can ignite the lubricant if there is too much in the barrel!

Note how the combustion chamber is reinforced with a couple of steel bands (these can be found at almost any hardware store). We have had combustion chambers split open when fired, and these steel bands definitely help to reinforce the chamber and reduce the chance of splitting.

Most of the remainder of the Howitzer is mainly for show and not functionality, but it does make a nice impression. Start with a piece of plywood, and simply stack 2- × 6-inch beams in a "staircase" fashion (here the 2 × 6's are stacked 10 high). Screw them together as you go, or band them all together with some strips of steel and screws, as in Figs. 19-9 and 19-10.

The HHO igniter and remote switch are attached to the end cap, used to fire the Howitzer (Fig. 19-10).

The rope is secured to the ground with stakes and helps to hold the HHO Howitzer in place because it does want to roll back from the impact when fired (Fig. 19-11).

This HHO Howitzer is no toy and can launch a projectile with considerable velocity. Even a large, lightweight, and smooth-faced piece of plastic can do some real damage from this howitzer (Figs. 19-12 and 19-13).

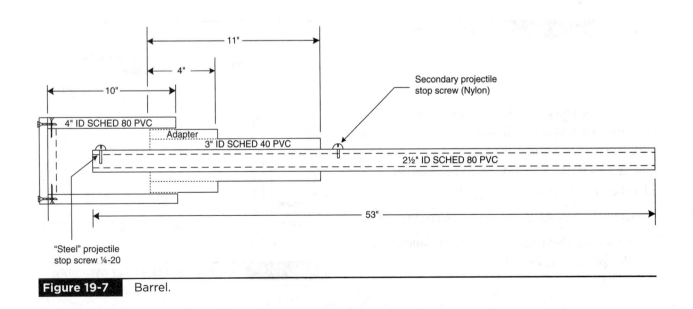

Figure 19-7   Barrel.

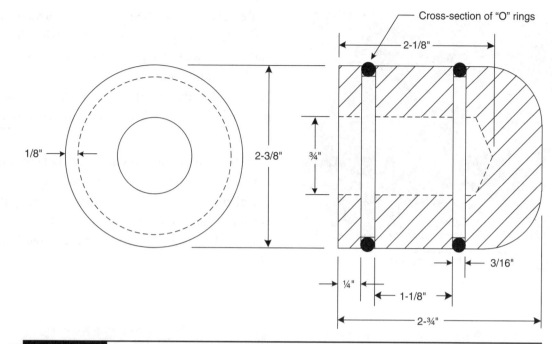

Figure 19-8   Projectile.

Figure 19-9   Main "body" of howitzer.

Figure 19-11   Front view.

Figure 19-10   Mount and breach.

Figure 19-12   Impact of the 50-millimeter projectile from 100 ft.

Figure 19-13   Impact, reverse side.

Be very careful when operating—always keep a safe distance (at least 10 or 15 ft) when firing, and always wear eye protection. This is not so loud to require hearing protection, but it does have a definitive report. As with a firearm, never point it at anyone or fire it when someone is down range.

Also be careful not to stand directly behind this device. We have had more than one combustion chamber fail on us during test firing—usually it is the end cap being blown off the back when too much HHO gas is released into the combustion chamber.

TABLE 19-1	Parts List
Quantity	**Description**
	53-inch × 2½-inch inner diameter tube PVC Sch 80
	5-inch × 10-inch × ½-inch thick PVC sheet
	11-inch × 3-inch inner diameter tube PVC Sch 40
	4-inch × 3-inch inner diameter tube PVC Sch 40 (also cut along length)
	10-inch × 4-inch inner diameter tube PVC Sch 80
4	Flathead screw, ¾-inch 6-32
4	Flathead screw, 1-inch 6-32
8	Flathead screw, 1¼-inch 6-32
2	2½-inch 3/16-inch diameter stainless screw
2	Electrical lug
	Ball valve ½-inch threads
	¼–20 set screw, metal
	¼–20 set screw, nylon
2	1-inch width metal straps for combustion chamber structural integrity support
	2-3/8-inch outer diameter plastic cylinder at least 3-inch length, solid core (this will be cut and crafted into the projectile)
2	2½-inch diameter 3/16-inch thickness rubber o-ring
	HHO Igniter (see Chap. 16)
	Optional ¾-inch thick 3½-ft × 1½-ft plywood base
7	Optional 2 × 6-inch × 8-ft wood beams for platform
2	Medium eyehooks (at least 1-inch diameter eyehook with ¼-inch diameter threads)
	10 ft of rope
	Optional wheels for platform
	Optional laser pointer mounted to barrel for alignment and aiming purposes

CHAPTER 20

Faraday Cage

Overview

A Faraday cage is used to shield against electromagnetic (EM) radiation. It is basically a metal box through which electromagnetic energy cannot pass or is greatly attenuated (the energy conducts around the outside of the box but does not pass through the interior). An ungrounded Faraday cage will still protect against EM radiation, but anyone touching the outside of the cage may receive a shock based on the amount of energy built up that has not yet dissipated. For this reason, a good ground connection to the Faraday cage is recommended to prevent painful and even dangerous shocks, especially when working with high-voltage items such as Tesla coils.

Hazards

Some cutting and hammering hazards during assembly, minimal for anyone familiar with carpentry.

Difficulty

Requires intermediate carpentry skills.

Tools

Basic carpentry tools such as a saw, hammer, drill, and square.

Assembly

The section shows how to construct a Faraday cage (Fig. 20-1) for attenuating the EM interference

(EMI) energy associated with a Tesla coil or similar apparatus. The device will provide excellent shielding of frequencies up to 300 MHz. When used with a line filter properly installed, interference should be reduced to a value below that which can cause interference with other electronic equipment.

1. Build a wooden frame around the coil operating area using standard building codes and dimensions. Note the size to allow proper operation of spark discharge and so on.

2. Obtain a suitable amount of hardware cloth with ½- × ½-inch squares and attach them to the frame members. Join pieces of hardware cloth by allowing to overlap 1 inch or two squares. Weave in No. 18 wire to secure sections. Solder

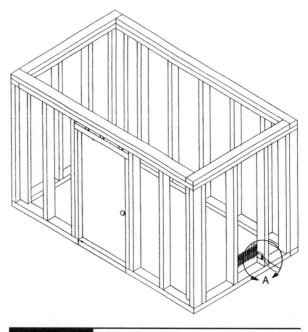

Figure 20-1   Basic Faraday cage.

every 1 inch or two squares. (Optionally, use standard ¼-inch-gap wire mesh available from local hardware stores. Hardware cloth is unnecessary. Overlap at seams about 2 inches, and secure to frame with heavy-duty staples (see Figs. 20-14 to 20-19).

3. The bottom and top of the cage may be solid sheet metal on similar hardware cloth. Pieces must be joined as shown to prevent EMI leakage.

4. Obtain an EMI line filter and install as shown. Note that there must be no conducting object or wires that can feed through cage without going through the filter. Control may be achieved by controlling alternating-current (ac) power externally before going through the filter.

First, build the frame: 1 × 3 lumber is fine to use because this is not a load-bearing structure. Get your materials (lumber and wire mesh) according to the size of the cage you are building. The illustrations herein are for a medium-sized walk-in cage, but you can, of course, build larger or smaller depending on your needs and available space (Figs. 20-2 through 20-10). Note that a radiofrequency interference (RFI) line filter also

should be connected to the screen. Earth ground via heavy wire to water pipe, direct-earth ground stake, and so on.

If you will be putting a Tesla coil inside the Faraday cage, allow sufficient clearance for discharge. Also, the cage may add capacitance to the system and cause a retuning of the Tesla coil to be required for optimal functioning.

Following is an example of the Faraday cage used in our shop. It is a simplified version of the plans shown here and works well. We have put this directly between two BTC70s firing full blast, and nobody has died yet.

First assemble the framework with the 1- × 3-inch beams, securing the corners with 6-inch triangles cut from the 4- × 8-ft piece of plywood. Also cut a door at 3 ft wide by 5 ft high (Fig. 20-10).

You will need 14 beams and 20 triangles. Secure them together with wood screws into a frame using the triangles and short-cut pieces of 1 × 3 to reinforce the corners, tops, and edges (Fig. 20-11). You can, of course, adjust the size (and the size of the securing edge pieces) to your preference.

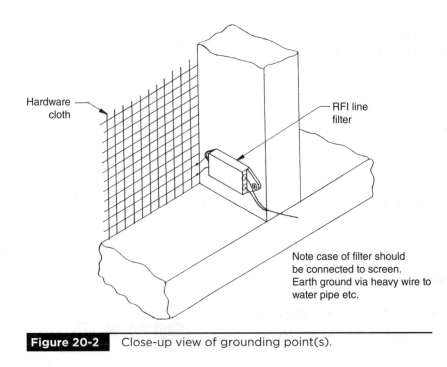

Hardware cloth

RFI line filter

Note case of filter should be connected to screen. Earth ground via heavy wire to water pipe etc.

Figure 20-2 Close-up view of grounding point(s).

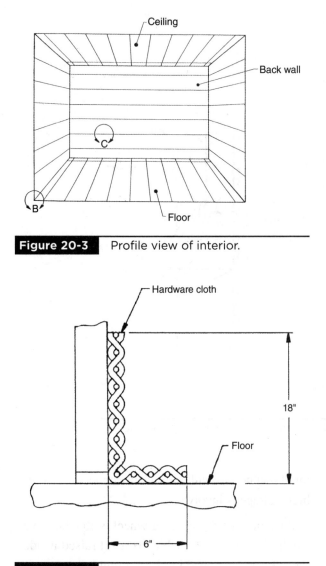

Figure 20-3   Profile view of interior.

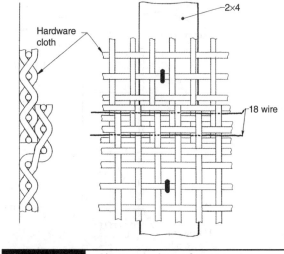

Figure 20-5   Close-up view of seam.

Figure 20-4   Close-up view of corner.

The corners should be reinforced on all sides with plywood triangles drilled into the 1×3 beams (Fig. 20-12).

With the finished frame, mount the front door from the large-cut section of plywood. This Faraday cage is not meant to be a sturdy construction, so attach a bracket as a brace at the corner of the door to hold it up; otherwise, it will pull the frame crooked over time. A slot down the middle and a couple of screws allow the height to be adjusted (Fig. 20-13).

Then completely cover the inside and door with wire mesh (but not the floor) (Fig. 20-14).

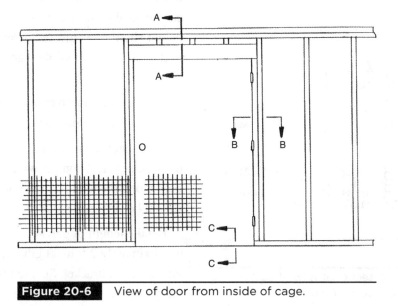

Figure 20-6   View of door from inside of cage.

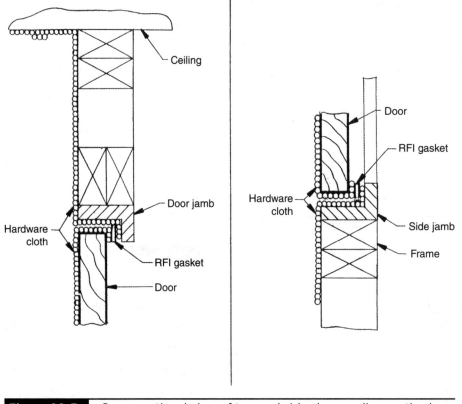

Figure 20-7 Cross-sectional view of top and side-door sealing method.

Figure 20-15 shows detail of how a wire-mesh strip electrically attaches the door to the cage. A short length of flexible high-voltage wire soldered to alligator clips also can be used to attach the door's wire mesh to the rest of the cage, or use one in addition to the wire mesh for the added peace of

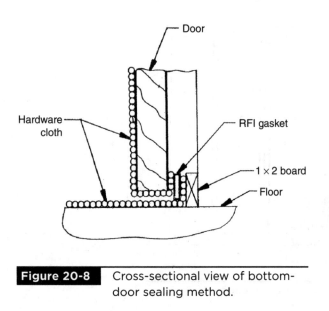

Figure 20-8 Cross-sectional view of bottom-door sealing method.

mind that comes with redundant connections when high voltage is involved.

Also note how the caster wheel is attached to a small plywood block, which is itself raised inside the cage and screwed in place through the outer frame. An L-bracket or another square of plywood can be used to secure the block vertically. The height should be adjusted to keep the cage as close to the ground as possible—about ¼ inch should do fine.

Another wire-mesh connection is made at the upper part of the door to the main cage (Fig. 20-16).

Note that the mesh on the door does not need to make physical contact with the mesh inside the cage along the top and at the opening side, but it should be very close.

Bend a length of galvanized steel or aluminum, approximately 2 ft in length and 6 inches wide, into a long L-bracket. Secure this to the outside of the door with short wood screws such that it will

Allow sufficient clearance for discharge when used for Tesla Coils. Room may add capacity to system and cause need to retune system.

Back wall lined with hardware cloth

Figure 20-9 General interior setup.

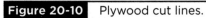

Figure 20-10 Plywood cut lines.

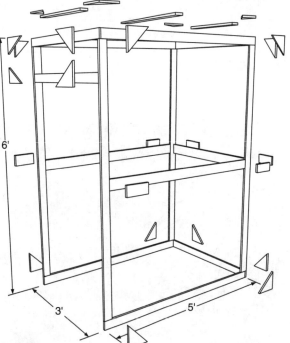

Figure 20-11 Faraday cage frame.

Figure 20-12 Top view of frame.

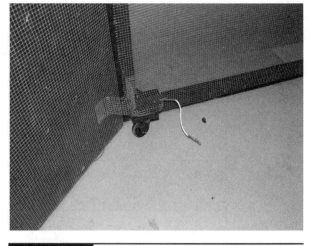

Figure 20-15 Door connection at bottom.

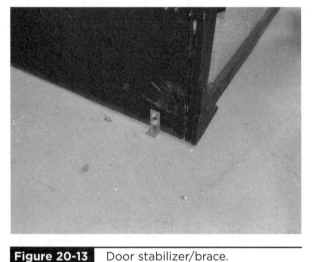

Figure 20-13 Door stabilizer/brace.

Figure 20-16 Door connection at the top.

Figure 20-14 Shop Faraday cage.

snugly hold the door closed. The length of this bracket helps to keep the door straight with the frame when it is closed. Drill a hole through the L-bracket and into the frame, and secure the door closed with a nail tethered to a rope (Fig. 20-17). It's inelegant, but it works. (A more elegant closing method, of course, can be devised if desired, but it should keep the door firmly closed and avoid any metal protruding into the cage past the wire mesh.) This also keeps the Faraday cage firmly closed when it is being used to ensure a physical continuity that keeps the electrical field on the outside of the cage. The nail also can be used

Figure 20-17   A long L-bracket secures the door to the frame.

Figure 20-19   Corner seam.

from the inside if there is no assistant to help with the cage, but ensure that the nail is pushed in completely flush because any metal protruding into the Faraday cage should be avoided (the inside should be a smooth surface of wire mesh on all sides to keep the electrical fields on the outside).

When attaching wire mesh to the inside of the cage, overlap the seams by an inch or so, securing in place every 2 or 3 inches with heavy-duty staples (Fig. 20-18).

Make sure that the corners overlap similarly (Fig. 20-19).

An uninsulated Faraday cage floor works fine for use with Tesla coils (Fig. 20-20) because any

stray arcs hitting the ground will be immediately grounded as long as the cage itself is attached to a grounding wire.

Attach the ground wire at about midheight to a nut and bolt going through the wire mesh with washers on both sides to hold it in place and establish a good connection with the wire mesh (Fig. 20-21).

Now put something in the cage! As you can see in Fig. 20-22, our "volunteer" is still mostly alive. (The sign on the door reads, "Caution: This Animal Bites." You might too after being sandwiched between 3 million volts of electricity.)

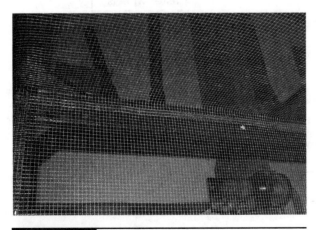

Figure 20-18   Overlapping seam.

Figure 20-20   Ground.

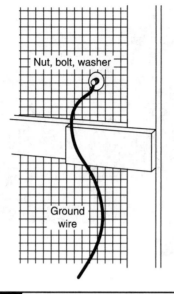

Figure 20-21 Ground wire connected to wire mesh.

Figure 20-22 Not dead yet.

The parts list for the Faraday cage will depend on the size of the cage that is being made. However, as a guide, a parts list is shown for the 3- × 5-ft Faraday cage shown here (the second cage in Figs. 20-14 through 20-22).

Note that the Faraday cage must control 2 types of radiation: 1) Radiated Emissions that travel through the atmosphere and interfere with radio waves, and 2) Conducted Emissions that travel through a structure, or have wiring, power lines, etc.

Larger constructions will of course require larger sized-components, as the cage shown is for small coils & people. A Faraday cage built to contain the Tesla coil of Chap. 9, for example, would be at a minimum 12-feet high and 20-feet square! It might be easier at this point to operate such a large Tesla coil inside a grounded metal garage (metal on all sides and roof), which would function just like a Faraday cage.

Quantity	Part
TABLE 20-1 Parts List	
1	Wire-mesh roll (or heavy-duty screen), 3 ft wide × 40 ft long
4	Caster wheels
2–3	Door hinges
1	Handle
1	10 ft of 16-gauge wire
4–6	Alligator clips
1	4- × 8-ft sheet of 3/8-inch plywood
14	1- × 3-inch (or 2- × 4-inch, if preferred) wood beams, 6 ft long
1	Heavy-duty stapler + heavy-duty staples
150	¾-inch sheetrock screws

CHAPTER 21

Test Supply

This is a test supply for 100- to 1000-W systems (Fig. 21-1).

Please read the following carefully:

This piece of equipment should be used when direct line voltage circuits are being tested. It is a valuable, useful, and possibly lifesaving addition to any electronics lab; it certainly is a circuit-saver!

Many of the higher-powered projects in this book use half and full bridge switching circuits. These circuits unfortunately can be hazardous because the common voltage line (ref. negative rail) is almost 200 V below earth ground. If you were to hook a ground lead such as a scope to this line, you would get a line short that could wipe out the test equipment and circuit under test.

One way to overcome this is to unground the test equipment. However, this can be deadly, especially on damp cement floors, and should be considered strictly forbidden.

A piece of test equipment that is recommended is shown and uses a 1-kW 120/120-Vac *isolation transformer*. Note that an auto transformer is not isolated and will only make the situation worse. **Beware of this when obtaining this crucial part!**

The addition of a variac is strongly recommended because this will allow slowly applying voltage to a virgin circuit rather than full voltage that can destroy the circuitry should there be a catastrophic fault.

Switchable ballast such as a light bulb also can provide a good degree of circuit protection when first hit with power.

A 300-Vac voltmeter and 10-amp alternating-current (ac) ammeter are advised.

An ac ground check is advised and can be a set of terminals for an ohmmeter that is set to verify leakage or miswires where the earth-ground green wire is somehow connected to the ac neutral (white) lead or ac hot (black) lead.

This piece of test equipment will make your effort 10 times safer and easier, so it is well worth the time and cost to build. All our technician and testing stations have these setups as described, and I can say that they have made testing circuits simple and safe, believe me!

Layout is not critical but should follow good and proper construction technique because it may save your life!

Note that the circuit in Fig. 21-1 is shown for up to 1 kW of usable bench power. You may scale up or down for more or less power.

> **NOTE** The ballast lamp should be rated at about the impedance of the unit under test when at full power.

Test Supply Operating Instructions

1. Turn the main power switch OFF.
2. Variac to zero.

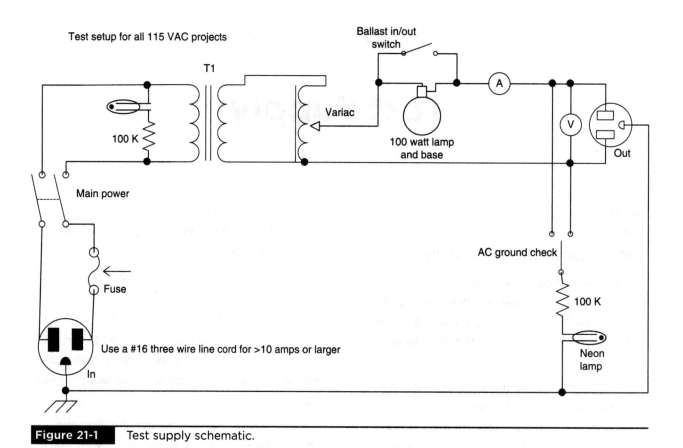

Figure 21-1   Test supply schematic.

3. Ballast switch IN.

4. Setup unit to test and plug in.

5. Turn ON the main power switch and slowly advance the variac to operating voltage; note ac current is as test procedure.

6. Check the ac ground switch. May glow slightly owing to capacitive feed through. Force an ac ground to get used to brightness.

Ballast Box (Fig. 21-2)

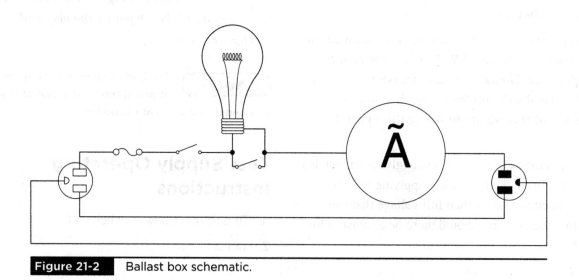

Figure 21-2   Ballast box schematic.

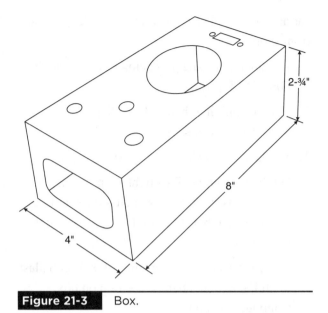

Figure 21-3   Box.

Figure 21-5   Bottom with feet.

Ballast Box Assembly

First, obtain or fabricate a box about 8 inches long × 4 inches wide × 2¾ inches high. Use a nonconducting material such as wood or plastic. Then cut holes for components. The placement and sizes don't need to be precise, but things should be sized and positioned according to the components being used (Fig. 21-3).

Next, install the components and wire them together. For some components, it works better to first secure them in the box before soldering/ wiring; other components should be soldered/ wired before they are secured in the box. Which method to use will become apparent as the box is assembled (Fig. 21-4).

With the components installed and wired, attach some rubber feet to a bottom sheet, and secure it to the box to hold the wiring inside (Fig. 21-5).

Put a standard 150-W incandescent light bulb in the socket, and the test supply is now ready to use (Fig. 21-6).

Ballast Box Operating Instructions

The ballast box is a simplified version of the test supply not using the isolation transformer or the

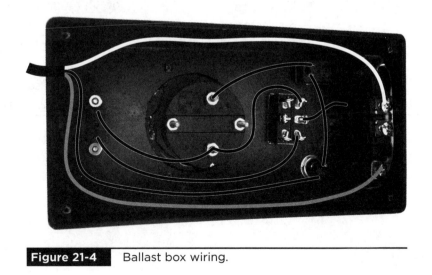

Figure 21-4   Ballast box wiring.

Figure 21-6 Ballast box.

variac, and it is used to prevent circuit destruction from shorts.

1. Be sure ballast and unit being tested are in the OFF/SAFE position.

2. Plug the unit into ballast box. Plug the ballast box into the wall socket.

3. Set the ballast box "green" position onbox (ON).

4. While watching ballast light, turn the unit on. If the ballast light turns on, either the unit has a short or the ballast light does not have high enough wattage.

5. If light stays off, turn the unit off and set ballast to OFF ("red" nonballast position on box). Continue operational check.

NOTE You may keep ballast box in the "green" ballast position for safety during the operational check, but be aware that the unit will not receive full current.

TABLE 21-1	Parts List
Quantity	**Description**
	4- × 8- × 2¾-inch plastic box
	4- × 8-inch plastic sheet 1/8-inch thickness
4	rubber feet
	electrical wall socket
	on/off toggle switch
	3-selector on-off-on (centerpost off) toggle switch
	fuse holder & fuse
	meter
	lightbulb socket
	several short runs of electrical wire (Fig. 21-4)

Index

Note: Page numbers followed by *f* denote figures; page numbers followed by *t* denote tables.